L. D. Kapoor, PhD

Opium Poppy: Botany, Chemistry, and Pharmacology

Pre-publication REVIEWS, COMMENTARIES, EVALUATIONS . . .

"The opium poppy *(papaver somniferum)* is said to be the oldest recognized medicinal plant in the world and is still very important, as it furnishes a most potent pain killer, morphine. In this monograph, Dr. Kapoor explains and develops in detail the importance of this hallowed drug. The book will be of incalculable value to anyone interested in this very important plant entity. It will be of most definite interest to the botanist, the plant physiologist, the phytochemist, the pharmacologist, the toxicologist, and the clinician. Every science library should have a copy on its shelves."

George M. Hocking, PhD
School of Pharmacy,
Auburn University, USA

Food Products Press
An Imprint of The Haworth Press, Inc.

Opium Poppy
Botany, Chemistry,
and Pharmacology

New, Recent, and Forthcoming Titles
of Related Interest
from *FOOD PRODUCTS PRESS*

Biodiversity and Pest Management in Agroecosystems by Miguel A. Altieri

Winemaking Basics by C. S. Ough

Statistical Methods for Food and Agriculture edited by Filmore E. Bender, Larry W. Douglass, and Amihud Kramer

The Highbush Blueberry and its Management by Robert E. Gough

Vintage Wine Book: A Practical Guide to the History of Wine, Winemaking, Classification, and Selection by the Sommelier Executive Council

Herbs of Choice: The Therapeutic Use of Phytomedicinals by Varro E. Tyler

Managing the Potato Production System by Bill E. Dean

Glossary of Vital Terms for the Home Gardener by Robert E. Gough

The Honest Herbal: A Sensible Guide to the Use of Herbs and Related Remedies by Varro E. Tyler

Seed Quality: Basic Mechanisms and Agricultural Implications by Amarjit S. Basra

Bramble Production: The Management and Marketing of Raspberries and Blackberries by Perry C. Crandall

Opium Poppy: Botany, Chemistry, and Pharmacology by L. D. Kapoor

Opium Poppy
Botany, Chemistry, and Pharmacology

L. D. Kapoor, PhD

Food Products Press
An Imprint of The Haworth Press, Inc.
New York • London

Published by

Food Products Press, an imprint of The Haworth Press, Inc., 10 Alice Street, Binghamton, NY 13904-1580

Library of Congress Cataloging-in-Publication Data

Kapoor, L. D.
    Opium poppy : botany, chemistry, and pharmacology / L.D. Kapoor.
      p.   cm.
    Includes bibliographical references and index.
    ISBN 1-56024-923-4 (acid-free paper)
    1. Opium. 2. Opium poppy. I. Title.
RS165.06K37   1995
615'.323122–dc20   94-20779
                                                     CIP

To
my wife,
Prakash

ABOUT THE AUTHOR

L. D. Kapoor, PhD, is a retired scientist from the National Botanical Research Institute, a national laboratory under the Council of Scientific and Industrial Research in Lucknow, India. Currently residing in New Jersey, Dr. Kapoor is consultant to a number of pharmaceutical concerns working on natural products. He has served as an advisor in Sri Lanka and Bhutan for the commercial production of ayurvedic drugs for pharmaceutical preparations, and was invited by the government of Nepal to advise on the establishment of herbal farms for the commercial production of herbs. A fellow of the Indian Academy of Science, Bangalore, and the American Society of Pharmacognosy, Dr. Kapoor is the author of four books and over 250 research papers.

CONTENTS

Foreword

One of the most meaningful contributions to the study of a subject is a book that reviews and summarizes the scientific literature, offering insight into the mechanics and seduction of the topic. Such a task requires diligence, patience, and considerable time, but enables others to gain a comprehensive understanding of the material. A well-written review provides a starting point for further exploration and makes available the basic information supply necessary for charting progress. Indeed, we are fortunate to have Dr. L. D. Kapoor, a noted scientist with expertise in medicinal species, to lead us in this book on a journey exploring the botany, chemistry, and pharmacology of the opium poppy.

The poppy, more than any other medicinal species, has been used and misused by humankind from the early days of history. Alkaloids isolated from this plant have played a significant pharmaceutical role in relieving pain, but have also captured the very being of individuals, leading them into that never-ending, painful anguish associated with narcotic dependency. Even today, the production and sale of this plant is politically charged. Grown as a narcotic, opium offers source families a means for making a living, but causes the receiving families to suffer. By studying and understanding the plant, we can apply the knowledge gained for the enhancement, not detriment of life.

Readers of this book will observe that the 11 chapters cover a variety of subjects, providing particulars on agronomy, applications, botany, chemistry, and history. Those using the book as a reference will note extensive details and numerous citations. Collected data are presented in tables and figures for summarization and clarification. If one opens to any page, some interesting and useful fact appears, enabling the reader to strengthen his or her understanding about the opium plant.

For the above reasons, I both congratulate and thank the author.

As a companion in the search for knowledge, I appreciate the effort that was necessary to accomplish the writing of this book. As a member of the community of scientists interested in medicinal species, I recognize the value of this distinctive volume. To the readers I say this timely work will undoubtedly prove valuable whether you are interested in the botany, chemistry, or pharmacology of opium.

Lyle E. Craker
Editor,
Journal of Herbs, Spices & Medicinal Plants;
Professor and Head,
Department of Plant and Soil Sciences,
University of Massachusetts, Amherst

Preface

Opium poppy (*Papaver somniferum* L.) is perhaps the earliest medicinal plant known to mankind. It was well known in ancient Greece, and Hippocrates mentions the use of poppy juice as cathartic, hypnotic, narcotic, and styptic. Pliny the Elder indicates the use of the seed as hypnotic and the latex for headaches, arthritis, and curing wounds. Dioscorides in A.D. 77 distinguished between the latex of the capsules and a whole plant extract. In time the cultivation of opium spread from Asia Minor, Egypt, to Persia, and on to India and China. The smoking of opium was first noted to be extensive in China and the Far East in the later part of the eighteenth century. Traffic and subsequent addiction proved to be useful sources of revenue.

In 1803, J. F. Dorosne isolated a mixture of morphine and narcotine, the first reported alkaloid from the opium (latex) which came from the capsule of the plant *Papaver somniferum*. A year later M. A. Seguin reported the isolation of morphine and a similar isolation was reported by F. W. Sertürner in 1806, but its structure was eventually deduced by J.M. Gulland and R. Robinson 120 years after the discovery of morphine. The latex, which is rich in alkaloids, is contained only in an anastomosing system of vessels or laticifers in the capsules where-from the bulk of exudate is recovered when incised. The laticifers occur in other parts of the plant, viz. leaf, stem, roots, petals, and sepals, but not in stamen or seeds.

As laticifers are found in phloem, their course and distribution can be followed by tracing the course of vascular bundles. The vascular bundles of stamens and ovules are devoid of laticifers, hence no latex and therefore no alkaloids. The laticifers are absent during the stage of life history covered by the pollen, ovules, seeds, and the young seedling.

Surprisingly, little work has been reported on the ontogeny, development, and distribution of laticifers. Prof. J. W. Fairbairn initiated the subcellular study of laticifers with a view to correlate the distribution of laticifers and presence of alkaloids in different parts of *Papaver somniferum*.

With the help of the electron microscope Fairbairn and his colleagues arrived at interesting results: that there are two types of alkaloidal vesicles, the latex one with smooth or slightly granulated walls and the other with a distinct cap which showed structural zonation. It has been suggested that the latter was the site of biosynthetic activity while the comparatively smooth vesicles may function for storage only.

Over 42 alkaloids have been isolated from opium poppy but only a few are of major significance. These are morphine, codeine, thebaine, BG, and papaverine, which have positive use in pharmaceutical preparations. The other alkaloids are either used as precursors or are intermediates in the pathways of biosynthesis.

With the advent of radioactive techniques after World War II, various schools of chemistry in the USA, Europe, and England have identified definite precursors and intermediates and pathways for the biosynthesis of the major alkaloids in the life history of the plant. However, our knowledge of enzymes responsible for the metabolism needs more precise information as to which specific enzyme is catalytic for the specific alkaloid. Isolation of biologically active enzyme preparations from alkaloid-producing cells faces enormous difficulties in practical experiments. Enzymatics of alkaloid metabolism have been investigated in detail only in a few microbial systems and cell cultures of a few higher plants (*Penicillium cyclopium, Catharanthus roseus*). The enzymology of biosynthetic transformations is still in its infancy, but from the growth and development of this branch of plant biochemistry will come major contributions in the future.

With the advancement of organic chemistry and sophisticated techniques, many alkaloids have been synthesized, but the major alkaloids of morphinan group are still derived from the vegetable source, viz. opium poppy (*Papaver somniferum*).

On the one hand the Morphinan alkaloids are a boon to alleviate the painful sufferings of human beings, but on the other hand they are a curse as narcotics to bring more miserable suffering and deterioration in human values. Drug abuse is a global problem and all the national governments are trying to control its illicit traffic, possession, and use. It would be too much to offer any alternative method of its control but it is suggested that illicit cultivation of *Papaver somniferum* must be stopped internationally and forcefully and cultivation of *Papaver bracteatum* as an alternative crop may be encouraged, which can supply the desired alkaloids for pharmaceutical purposes.

In this book I discuss the botanical aspects, viz. botany, agronomy, anatomy, physiology, and chemical aspects of major alkaloids, their biosynthesis, the precursors identified, and other intermediary alkaloids all in one volume. It will serve as a useful companion volume to the students of pharmacy, botany, and chemistry medicines, and pharmacology for overall better understanding of opium poppy.

The history of opium poppy is traced from the ancient civilization through the present day in the introductory chapter. Studies in botany including cytogenetics, plant physiology, and agronomy are described in the next few chapters. Developmental anatomy of the capsule with special reference to laticiferous vessels, the repository of latex, is described and illustrated along with their presence or absence in the flowering parts in Chapter 6. The constitution of amphicribral bundles and their dissolution is depicted in Chapter 7. The chemistry of opium alkaloids and their biosynthesis is discussed in Chapters 8 and 9. The site of synthesis and the role of alkaloids is briefly discussed in Chapter 10. Evaluation of analgesic actions of morphine in various pain models in experimental animals is discussed in the last chapter. For the convenience of readers, references are given at the end of each chapter.

Frequent references are made to the outstanding treatise, "Introduction to Alkaloids," by Geoffrey A. Cordell, "Alkaloid Biology and Metabolism in Plants," by George R. Waller and Edmund K. Nowacki, "The Isoquinoline Alkaloids Chemistry and Pharmacology," by Maurice Shamma, and "Plant Anatomy," by K. Esau. Publications of the late Prof. J. W. Fairbairn and Prof. E. Brochmann-Hanssen have been very helpful to compile this volume. For more detailed information the readers are advised to refer to the original papers.

Acknowledgements

I owe my gratitude to the late Sir R. N. Chopra, an authority on drug addiction in India who was a source of inspiration to undertake this subject matter. I am equally grateful to the late Prof. J. W. Fairbairn for initiating his research on opium poppy with me, to start with, and for his invaluable guidance during the experimental phase. Thanks are due to Dr. P. N. Suwal, the late Dr. S. B. Challen, and the technical staff, Department of Pharmacognosy, London School of Pharmacy, University of London for their assistance.

My special thanks to Dr. James A. Duke, Prof. G. M. Hocking, Dr. I. C. Chopra, Dr. D. P. Jhamb, Prof. B. K. Nayer, Dr. A. B. Segalman, Dr. J. D. Phillipson, Prof. Lyle E. Craker, Dr. Bal K. Kaul, Dr. K. R. Khanna, Dr. Rajinder Gupta, Mr. B. M. Kapur, Mr. Himanshu Juneja, Mr. Rahul Thukral, and Dr. V. Chandra, who have been helpful in many ways.

For their moral support and cooperation I am obliged to Suman, Anil, Dr. Kusum, Ashish, Sita, and Aparna in helping me to complete this assignment. Special grateful thanks to Mr. Bill Cohen and his colleagues at The Haworth Press, Inc. for making the publication of this volume possible.

My sincere thanks are extended to University of London, Oxford University Press, England, John Wiley and Sons, Inc., Plenum Press, New York, Academic Press, Inc., Pergamon Press, Inc., George Thieme Verlag, University of Chicago Press, Macmillan Magazines, Ltd., American Chemical Society, Chief Editor Wealth of India, and Director of Central Institute of Medicinal and Aromatic plants, Lucknow, for permission to reproduce extracts, abstracts, tables, or figures from their publications. Each item is acknowledged individually in the text.

I am particularly thankful to Dr. E. Brochmann-Hanssen, Department of Pharmaceutical Chemistry, University of California, San Francisco for permission to abstract or reproduce text and figures from his masterly work on the biosynthesis of morphinan alkaloids. For photographic help, I am grateful to Mr. Sanjiv Chadha.

I gratefully acknowledge the helpful cooperation received from the librarian and the staff of the Library of Science and Medicine, Rutgers University, New Jersey, USA.

I offer my sincere thanks to Mrs. Carol Ann Tucker of C.A.T.'s Typing Service, Dayton, New Jersey, for her efficient secretarial help, without which this project could not be finalized. Last but not least, I am indebted to the tiny angels, Nakul, Kern, Ankit, and Anjali, who by their innocent activities provided the much-needed recreation and relaxation during the course of preparation of the manuscript.

Dr. K. Ramabadran and Dr. M. Bansinath have very kindly contributed one chapter entitled "Evaluation of Analgesic Actions of Morphine in Various Pain Models in Experimental Animals" to this book, for which I am very grateful.

I am specially indebted to Dr. Jatin P. Shah, M.D., F.A.C.S. Chief, Head and Neck service, Memorial Sloan-Kettering Cancer Center, New York, for extending the lease of my life to complete the final phase of this book.

L. D. Kapoor

Chapter 1

Introduction

Opium poppy (*Papaver somniferum*) is one of the earliest medicinal plants known to mankind. The drug had its origin not in China where it was extensively produced and most widely used as thought earlier, but with some other yet earlier civilization.

The time and place of discovery of opium cannot be fixed accurately. Many authors have tried to trace the original habitat of opium poppy and opium. Though they differ in their approach, they agree on its origin in the remote past. Among relics of the Stone Age lake dwellers in Switzerland, about 4,000 years ago, the seeds and large seedheads of the cultivated poppy have been found. The most natural assumption is that they were grown for opium and not for oil which could be obtained from many other vegetable resources. Scott,[1] however, came to the conclusion that in Europe, opium was being taken before men were literate.

Opium poppy, perhaps, originated in Asia Minor.[40] The first and the most ancient testimony concerning the poppy is given in a small tablet of white clay found at Nippur during the excavations undertaken by the University of Pennsylvania Archaeological Mission.[2]

Nippur was the spiritual center of the Sumerians and lies to the south of modern Baghdad. The tablet is inscribed in Cuneiform script. Sumerians lived in the areas between the Tigris and Euphrates rivers in the lower half of present-day Iraq. They flourished from the fourth to third millenium B.C. and developed great proficiency in the arts, agriculture, and sciences in general. The earliest known mention of poppy is in the language of the Sumerians, a non-Semitic people who descended from the uplands of central Asia into Southern Mesopotamia, there founding a kingdom five or six thousand years before the birth of Christ.[3]

Anslinger and Tompkins[4] reported that the clay tablets of Sumerians displayed in writing how the juice of poppy was collected very early in the morning. This indicated that the Sumerians cultivated the poppy to extract opium about B.C. 3000 and they called it "Gil," which means happiness; this term is still used for opium in certain parts of the world. Dougherty[5] opined that opium must have been known to the Sumerians because they had an ideogram "Gil Hul" (as the joy plant), referring to opium poppy.

The Sumerians were succeeded by the Assyrians who also had a word for the juice of the poppy plant, namely "arat-pa-pal." It may well be that the latin word "Papaver" is derived from this etymological origin. The following description of preparation of opium appears in Thompson's[6] translation of the Assyrian medical tablets of about B.C. 700. "Early in the morning, old women, boys and girls collected the juice, scraping it off the notches (the poppy capsule) with an iron scoop and depositing the whole in an earthen pot." Thompson added, "It seems that nothing has changed in the method of collecting opium."

The Persians in their conquest of Assyria and Babylonia probably learned of opium from the people of those countries. It is in the sixth century B.C. that opium is mentioned in a Persian text. Cultivation of opium poppy was very ancient and opium was called *theriac, malideh,* or *afiun* by the Persians. The failure to mention opium in any of the Persian texts until the sixth century A.D. is explained by the fact that early Persians left no literature other than their own sacred books.[3]

The poppy was well known and the use of opium was popular in ancient Egypt. A. Fields[7] in Merk Report observed that it is from Egyptian Thebes that are gleaned such terms as opium thebaicum and the name of the alkaloid Thebaine. Galen[8] (ca. A.D. 130-200) affirmed that even Thoth, the bird-headed Egyptian god of letters, invention, and wisdom, who in later years was called Hermes Trismegistos, taught the mortals about the preparation of opium. Thoth was also the mouthpiece and recorder for other gods as well as the arbitrator of their disputes.

In the Ebers papyrus (ca. B.C. 1553 or 1550), a medical text, poppy capsules are mentioned along with other remedies. For example, an item appears, "Remedy to stop a crying child: Pods of poppy plant, fly dirt which is on the wall, make it into one, strain and take for four days, it acts at once."[9]

It appears that the use of opium in early Egypt was closely related to religious cults. The knowledge and use of the drug was an exclusive privilege of the priests, magicians, and warriors. The use of opium for euphoric purposes by the masses is described as secularization of a sacred plant.[10]

Earlier writers are of the opinion that the poppy and its cultivation were not known to primitive Egyptians. According to Woenig,[11] the ancient monuments or wall inscriptions of the earliest temples of those people did not show any flower, fruit, or seed. Schweinfurth[12] is, however, of the opinion that poppy was introduced and cultivated in Egypt shortly before the Roman era.

Poppy is referred to as Shepenn, the capsule as Sheppen, and the flowers as Shepenndsr. Gabra[13] observed that poppy products were used over the centuries and the term opium is found in the Greco-Roman period. Opium was used as a pain reliever and narcotic. It was a famous drug of Egypt and its specific preparations were concentrated in specific districts.

The Arabs are credited with the dissemination of knowledge to various parts of the world. Cultivation of opium poppy spread throughout the Arab Empire during the seventh century A.D.

Avicenna or Abu-Ali-Ibn-Sina (ca. A.D. 980-1037), the most celebrated Arabian philosopher and physician, wrote a thesis on opium, calling it the most powerful of narcotics and recommending it for diarrhea and diseases of the eyes. Avicenna became the father of "soothing syrups" by advising "If the child does not sleep properly mix some poppy with his food." Emboden[14] records that Avicenna died of opium intoxication in Persia in the year 1037 A.D. Ibin Sina, who lived in Bokhara during 980 A.D., stated that opium was widely used as a narcotic by the people of Khorasan and Bokhara. The use of opium as a narcotic was first started in Syria where it was eaten and added to sweets and spices. Arabs used to call opium poppy "Abou-el-noum" meaning "father of sleep." Cultivation of the poppy in Egypt developed during the Arab rule in the seventh century when it was cultivated in Alexandria and was exported not only to Europe but to India also.

Maimonides (1135-1204 A.D.), a great rabbi, Jewish scholar, and writer of the Middle Ages, and Averroes (ca. A.D. 1126-1198), a distinguished Spanish-Arabian physician and commentator on Aristotle, each wrote a thesis on the medical uses of opium.[15]

Opium was brought to China by the Arabs.[16] Whab-abi-Kabacha visited China during the later T'ang Dynasty (A.D. 618-907) and obtained permission to build a mosque at Canton. During the fourteenth century, Ibn Batuta (A.D. 1304-1377), properly named as Abu Abdullah Mohammed, the Arabian traveller, gave an account of his visit to China. Muslim influence, including the use of opium is still apparent in parts of China, especially in some of the communities in Southern Yunan, visited by this author. Dr. Eldein[17] records that poppy is mentioned in a medical book of the tenth century as *minang* (i.e., millet vessel) and *yingsu* (i.e., jar millet). In the medical works compiled by Sung during 1057 A.D. poppy is shown as cultivated all over China as an ornamental or food plant. The Chinese gained the knowledge of medicinal uses of poppy from Arab traders in the thirteenth century. By the fifteenth century, the cultivation of poppy was well established as a food plant, but opium was still imported from India.[18]

Tracing the history of poppy in antiquity, Kritikos and Papadaki[2] have made a study in depth of classical texts and archaeological finds in Greece and concluded that the poppy plant and its hypnotic properties were well known in the classical period of ancient Greece and are mentioned by the contemporary writers. The first written record of poppy is found in Hesiod[19] (eighth century B.C.), who states that in the vicinity of Corinth there was a city named McKone (Poppy-town). This city perhaps received its name from the extensive cultivation of the poppy in the area. It was regarded as a magical and also as a poisonous plant and was used in religious ceremonies. Its therapeutic use came a little later. The ancient Greeks portrayed the divinities Hypnos (Sleep), Nyx (Night), and Thanatos (death) wreathed with poppies or carrying poppies in their hands. They adorned statues of Apollo, Asklepios, Pluto, Demeter, Aphrodite, Kybele, Isis, and other dieties in like manner. Sometimes ears of corn were added to the bunch of poppies.

It has been mentioned in Greek mythology that Demeter, in despair over the seizure of her daughter Persephone by Pluto, ate poppies with a view to fall asleep and forget her grief, and accordingly she offered poppy to Triptolemus also in order to induce sleep.[20]

The word opium seems to be of Greek origin, from opos–a juice. Opium probably was brought into Greece from Asia Minor.[20] Homer mentions in the *Iliad*,[21] "The poppy which in the garden is weighed down by fruit and vernal showers, droops its head on one side." He also mentions the Cup of Helen as "inducing forgetfulness of pain and sense of evil." He also spoke of the intoxicating poppy "saturated with lethal slumber."[22]

Hippocrates[23] (B.C. 460-359) recommended drinking the juice of white poppy mixed with the seeds of the nettle for leukorrhea. He also recommended the poppy for "uterine suffocation" and describes the juice as "hypnotic meconium." According to him, Heraclitus (B.C. 641-575), the emperor of Byzantine, founded the empiric school of medicine based on the tripod of observation, history, and analogy. He rejected anatomy as valueless and depended on opium as an anodyne. Theophrastus[24] (ca. B.C. 372-257) also referred to the milky juice of poppy known as meconium and described the method of collecting it. Dioscorides[25] (ca. A.D. 50) described opium as "pain easer and sleep causer." He distinguished between the latex of the capsule and an extract of the whole plant. Hippocrates[23] makes frequent mention of poppy as being used in medicinal preparations. Aristotle[26] (B.C. 384-322) has mentioned poppy as a hypnotic drug. Galen[8] has stated "Opium is the strongest of the drugs which numbs the senses and induces deadening sleep."

Alex Tschirch[27] places the use of opium in B.C. third and fourth centuries and states that Roman physicians imported this drug from Greece or Greek colonies. After studying all the available evidences, Kritikos and Papadaki[2] concluded that poppy was known from the earliest times in Greece and is mentioned in the earliest records authored by Homer, Hesiod et al. From the time of Pliny,[28] the juice of the capsule was known as opium and the leaves and fruit as mekonion, but opium was considered more potent than mekonion. The use of opium by nasal inhalation as a hypnotic drug was the most suitable method of inducing sleep. It was also used for internal complaints by oral consumption and for external application by means of poultice and eyewashes, etc.

With the destruction of Corinth (B.C. 146), Greek medicine and

Greek physicians migrated to Rome. Many of the outstanding physicians of Rome were of Greek birth or descent, such as Antyllus, famous Greek Physician and surgeon of the second century A.D. and Saint Luke the Evangelist. Pedonius Dioscori or Dioscorides[2] (ca. A.D. 50), a Greek medical writer who served in Nero's Army wrote in his famed "De Universa Medicina" of opium as a "pain easer and a sleep causer." He distinguished between the latex of the capsules, "opes," and an extract of the whole plant. Scribenius Largus[2] (ca. A.D. 40), another Roman authority, describes a method of procuring opium from the capsules. The poppy is represented in Roman antique art as a mythological symbol of sleep and often represents the dispenser of sleep. Among the early Romans, the "poppy head" belonged to the mysteries of "Ceres" from old Italian mythology, the goddess of grain and harvest because she took papaver to "forget the pain" (ad obleveinem doloris). That is why a small earthen statue of Ceres-Iris with a torch holds poppy heads in her hand. Everywhere in antique art we meet with the poppy as a mythological symbol of sleep and even personification of the dispenser of sleep. Somnus in Roman mythology is the personification and god of sleep. The Greek Hypnos, a brother of Death (Mors or Thanatos) and a son of Night (Nox), is represented as a bearded man leaning over the sleeper and pouring on his eyelids the poppy juice contained in a vessel or horn which he holds in his hand. At a later date Somnus is depicted as a young genie carrying poppies and an opium horn with poppy stalks in his hand.[29]

The Roman poet Ovid (B.C. 43-A.D. 18) also recognized the narcotic properties of opium. In his *Fasti*, a poetical Roman Calendar, Ovid wrote, "Her calm brow wreathed with poppies night drew on and in her brain brought darkling dreams." Elsewhere he recited, "There are drugs which reduce the deep slumber and steep the vanquished eyes in Lethean night." Lucin (ca. A.D. 120-200), the celebrated Greek sophist and satirist, refers to poppy and associates it with sleep. In his true history, he comes to the Isle of Dreams where he finds the city of sleep "round about environed with a wood, the trees whereof are exceeding high poppies and mandragoras in which infinite numbers of owls do nestle. As one enters the city, on the right is the temple of Night, on the left is the Palace of

Sleep." In one of his poems on agriculture (Georgics) Virgil (B.C. 70-19), a famous Roman idyllic poet, has a beautiful phrase about the poppy: *Lethaco perfusa papavero somno* (poppies steeped in Lethe's slumber) and again in the Aeneid: *Spargens humida melle sporiferumque papaveri*. Perhaps Virgil refers to the use of opium when he relates how Cerberus, in Greek mythology, the dreaded three- headed dog with serpent's tail which guarded the entrance to Hades, was put to sleep by a drug.

It is mentioned in the Bible that "the Lord God caused a deep sleep to fall upon Adam and he slept: and He took one of his ribs and closed up the flesh instead thereof" (Genesis 2:21). This is one of the earliest records of surgical anesthesia and the first report of a thoracoplasty.[30]

According to some Jewish authorities (Haupt, Post), there is direct reference to poppy juice in the Bible as well as in the Talmud. The word rosh is mentioned in connection with the word La'anah (Jer. 8:14, 9:14, etc.). La'anah means "wormwood" or "absinthe" and rosch is translated in the authorized versions as "hemlock." However, rosch in Hebrew also means "head" and may refer to poppy head. Merosh or the juice of rosh translated as "poison water" may refer to the juice of the poppy. It is of interest that the Latin for poppy head is *"caputa"* which also means "head."[31]

The Samme de shinta of the Bible, like Helen's nepenthe, the "bhan" of the Arabian nights, and the "drowsy syrups" of Shakespeare were all probably combinations of opium, cannabis, mandrake, dewtry *(Datura stramonium)*, henbane *(Hyoscyamus)*, and hemlock *(Conium)*. The references to opium in the Talmud (A.D. 400) were probably borrowed from the Greek (Palestinian Talmud Ir. Abodah Zarah, 40a). It is likely that Greek and Jewish physicians exchanged ideas during Alexander's Indian expedition (ca. B.C. 330). Some writers maintain that even Aristotle (ca. B.C. 384-322), who lived about that time, received many ideas from the Jews.[28] Undoubtedly the use of opium along with many other drugs was acquired by the Jews during the periods of captivation in Assyria and in Babylon. Like the Babylonians and other ancient people, the early Jews considered disease as an expression of the wrath of God, to be removed by prayers, sacrifices, and moral reforms. At a later date, there occurred a separation of functions of priests and physicians, the latter utilizing drugs, diets, and surgery in the treatment of disease.[31]

Kritikos and Papadaki[2] observed that the Hebrews must have known about the use of poppy, as it occurred in their neighborhood. According to Marti,[32] the flora of that area included *Papaver somniferum* or *P. setigerum* (seed and oil). Rindley[33] mentions rosch (head) in the Bible, which is interpreted as a capsule of *Papaver setigerum*. According to him this was the gall mixed with vinegar which the Hebrews gave to Christ on the cross in order to alleviate His suffering. Walker[34] observed that the gall added to vinegar was the juice of poppy called *rosch* in Hebrew.

Discussing the archaeological evidence of opium poppy Bisset[44] observed that the capsules from the Cueva de los Murcielagos (Cave of Bats) near Albunol in the southern province of Granada, Spain, date from the Late Neolithic ca. 2500 B.C. and have been identified as being of *P. somniferum* L. The poppy finds from the rhineland region, from lakeside dwellings in France, Italy, and Switzerland and from Poland, some of which have been reported as *P. setigerum* are far from the Mediterranean distribution range of that species and may therefore have come from cultivated plants or from weeds among other cultivated plants. The massive finds of seeds in Germany and Switzerland suggest its use as a food or source of oil rather than as a medicine.

Krikorian[46] stated that no poppy material has been detected at any neolithic site in the eastern Mediterranean or near East. It is not until the Bronze age that knowledge of poppy and its products appears in the eastern Mediterranean. Perhaps the earliest indication recognized so far is a portrayal of the plant, including capsule at the palace of Knossos in Crete; it may belong to the latter part of the Middle Minoan III period, ca. 1600 B.C. Kritikos and Papadaki[2] reported about a remarkable terra-cotta figure of a goddess adorned with three scarified poppy capsules of *P. somniferum,* which have been recovered from a site at Gazi, west of Heraklion, also in Crete. This figure is of the Late Minoan III period ca. 1400-1350 B.C. The circumstances of the find and the associated artifacts, as well as finds elsewhere on the island, lead to the conclusion that during this period the use of opium for religious and probably also medicinal purposes was known on the island.

Contact between Crete and Cyprus goes back to at least the sixteenth century B.C. and recent work in northwestern Cyprus suggests that Cretan traders may have brought the opium poppy to the island at about that time or a little earlier.[47] From a study of the opium trade in the eastern Mediterranean it is observed that small long-necked juglets manufactured in Cyprus resembling a poppy capsule on the stem and which have been discovered in Egypt and elsewhere were used for the transport of opium.

These imported juglets appear in Egypt during the first half of XVIIIth dynasty, ca. 1567-1320 B.C. Other artifacts, faience, beads, earrings, and temple decorations show that it was during this period that Egyptians became acquainted with the plant itself.

In the tomb of the royal architect Cha who died during the reign of Amenophis (Amenhotep) III (ca. 1417-1379 B.C.) seven small alabaster vases were found including one similar in shape to the Cypriot juglets. One of these vases contained a product which according to chemical test was comprised of vegetable fat, together with iron and morphine.[50,51] Later chemical studies on the contents of cypriot juglets have implied the presence of opium in some cases but not in others.[49] Brooches have been recovered dating from the sixteenth century B.C. and having the characteristic feature of a grooved poppy capsule and its stem in Mycenae. It is also suggested that certain containers known as "aryballoi," usually meant for perfumes, might have held opium preparations.

The existence of poppy in Anatolia during the time of the Hittite Empire, ca. 1400-1200 B.C. is evidenced by two identical gold pins. Those faithfully reproduce details of the poppy capsule and its stalk down to a loose object inside (presumably representing seeds) so that the pins have a realistic rattle.

On the basis of botanical evidence, both Vesselovskaya[52] and Schultze-Motel[45] indicate that the Western Mediterranean must be recognized as the region where the poppy originated. According to Hammer and Fritsch[53] the present day distribution of *P. setigerum* (*P. somniferum* subsp setigerum) now relatively rare, comprises the northern part of the Mediterranean with the center of gravity in the West ranging from the Canary Islands as far as Greece and Cyprus.

In the taxanomic classification proposed by Vesslovskaya[52] there are six subspecies, but *P. setigerum* is excluded. The more recent one of Hammer[59] divides *P. somniferum* into three subspecies; subsp *somniferum* and subsp *songaricum* Basil, two cultivated races which differ in geographical distribution and in features of the stigma rays, and subsp *setigerum* (DC) Corb., the putative ancestral wild race. The two cultivated subspecies are each further divided into two convarities, one with indehiscent and the other with dehiscent capsules. Each covariety is in turn divided into 13 varieties on the basis of seed and flower colors as originally proposed by Danest.[55]

References to opium are not available in the classical Hindu literature in India, viz. Vedas, Puranas, and early medical classics, such as Charaka Samhita (third to second century B.C.), Sushruta Samhita (fifth to fourth century B.C.), and the works of Vagbhata (fifth century A.D.) viz. Ashtanga Hridaya and Ashtanga Samgraha. Dwarkanath[35] reported that references to opium appear in medical works belonging to the twelfth and thirteenth centuries A.D. onward. Ashwavaidyaka, a medical treatise written in twelfth and thirteenth centuries A.D. by Jayaditya, makes a reference to opium. Authorative Ayurvedic works on materia medica such as Dhanwantari nighantu (eighth century A.D.), Madanpala nighantu (1374 A.D.), and Raja Nighantu (1450 A.D.) have described the properties, actions, and indications of both cannabis and opium. According to Sharma,[36] medicinal preparations are described in Shodal Gadanigraha (ca. A.D. 1200) and Sharangdhar Samhita (1400 A.D.), where opium's use is prescribed in diarrhea and sexual debility.

Bhavamishra (fifteenth century A.D.), a contemporary of Paracelsus, has in his book *Bhavaprakasha* described the properties, actions, indications, and formulations of both cannabis and opium. Much later Ayurvedic medical works have given increasing importance to these two drugs.

It is, however, considered that opium was introduced in India by the Arabs who conquered Spain, Egypt, Asia Minor, Turkistan, Persia, and parts of India in the seventh century A.D. Many authors[18] believe that opium was introduced by Alexander the Great in 300 B.C.

Dwarkanath[35] opines that cannabis and opium have also been employed by traditional folk medicine in the treatment of diseases even as early as the fourth to third century B.C. During the last two centuries, traditional folk medicine and classical Indian medicine have become almost synonymous. Many drugs including those containing cannabis and opium have entered into the practice of classical Ayurveda.

Kritikos and Papadaki[2] are of the opinion that the information gathered from texts to the effect that opium was not known in India before the seventh century A.D. does not preclude the possibility of its having been known earlier, since Greek medicine in the third century B.C. had acquired many Indian drugs and conversely many Greek drugs had been introduced to Indian physicians. Thus, opium, which in the time of Hippocrates was known both in Greece and in Egypt, might well have reached India as well.

From the time of the Mogul conquest there appears the word Khas-Kish, referring to poppy seed, and Khas-Khas-rasa, the juice of the poppy. In modern Sanskrit the names abhena (foam) and abiphena (serpent's foam) are applied to opium.

The earliest mention of the cultivation of opium poppy in India is by the traveller Duarte Barbosa[43] in his descriptions of the Malabar coast in 1511 A.D. Mention has been made about Cambay and Malwa as the places where it was grown. It appears to have been cultivated first along the seacoast areas and to have penetrated later into the interior of the peninsula.[38]

The Portuguese historian Pyres in a letter to the King of Portugal in 1516 A.D. speaks of opium of Egypt and Bengal. In the Ain-l-Akbari compiled by Sheikh Abdul Fazal about 1590 A.D. the poppy is mentioned as a staple crop of the spring harvest of the then Subas of Agra, Oudh, and Allahabad.[37]

In the times of Moguls, the extensive cultivation of opium poppy made it an important article of trade with China and other Eastern countries. During the reign of Akbar, opium was made a state monopoly. It was grown in Oudh, Agra, Allahabad, Bengal, Orissa, and Bihar in India.

After the fall of the Mogul Empire, the state lost its hold on monopoly and control over the production and sale of opium was appropriated by a ring of merchants in Patna.[38] In A.D. 1757, the monopoly of the cultivation of poppy passed into the hands of the East India Company which had by that time assumed the responsibility for the collection of revenues in Bengal and Bihar.

Warren Hastings as Governor General brought the whole of opium trade under the control of the East India Company.

The cultivation and production of opium poppy was confined to three centers:

1. Patna or Bengal opium from poppy grown in Bihar and Bengal.
2. Benaras opium from United Provinces.
3. Malwa opium-produced in Rajputana, Gwaliar Bhopal, and Baroda.

The opium monopoly was promulgated in India by the Opium Act of 1857, the year when the British Government took over the administration from the East India Company.

The Central Government of India inherited this monopoly after independence in the year 1947 by virtue of which the cultivation and collection of opium, and manufacturing of opium alkaloids is controlled by Government.[37]

The use of opium in trade became a major factor in the spread of addiction throughout the world and especially in China. The British attitude can be judged by Lord Warren Hastings' pronouncement of opium as a "pernicious article of luxury which should not be permitted but for the purpose of foreign commerce only."[38]

Opium in Europe. Opium was first introduced in Europe in disguised form in various concoctions and confections containing numerous useless ingredients. Paracelsus Phillippus Aureolus (originally Theophrastus Bombastus Von Hohenheim 1493-1541 A.D.) has been referred to by Garrison as the "most original thinker of the sixteenth century." Perhaps he owed much of his therapeutic success to the administration of opium, "the stone of immortality." He wrote, "I have an arcanum which is called Laudanum, which is superior to everything when death is to be cheated."

He probably applied the name "laudanum" to several preparations all of which contained opium. The name may have been derived from the Latin, *Laudandum* ("something to be praised").[39]

For more than a century after Paracelsus, the term laudanum was applied to solid preparations, some containing opium, others of a purely mineral origin.

The first few editions of the London Pharmacopea contained

varying formulas for solid preparations of opium under the name of laudanum. This preparation contained opium, saffron, caster, diambra, ambergris, musk, and the oil of nutmeg, all of which first were to be made into a tincture which later was evaporated to form a pill mass.[39]

The alcoholic tincture known by the name of laudanum owes much of its origin to the celebrated seventeenth century English physician Thomas Sydenham,(1624-1689), nicknamed as "English Hippocrates." Sydenham's preparations were made from opium and saffron, extracted with canary wine, and a corresponding preparation was official in many pharmacopeas and is presented in the National Formulary by tincture opicrocata. In 1669, Sydenham described it as good for plague. Since he is reported to have fled from London during the Great Plague of 1665 along with most of other physicians of the city, his knowledge was probably secondhand and is derived from that of apothecaries who remained in London during the epidemic.

Sydenham's own opinion of the preparation is shown in the following quotation from his works. "I don't believe that this preparation has more virtues than the solid laudanum of the shops but it is more convenient to administer. Of all the remedies which a kind Providence has bestowed upon mankind for the purpose of lighting its miseries, there is not one which equals opium in its power to moderate the violence of so many maladies and even to cure some of them. Medicine would be a one arm-man if it did not possess its remedy. Laudanum is the best of all cordials; indeed it is the only genuine cordial that we possess today."

Another opium preparation no longer used is the vinegar of opium, Lancaster or Auaker's Black Drop. It originally was made three times the strength of laudanum by a complicated method based on several months' fermentation of wild crab apples with yeast.[39]

Quacks made Dover's powder famous. The name of Thomas Dover (1660-1742), an English physician, survives in Dover's powder (Pulvis ipecacunhe opuii). It was not generally known that he was a devoted servant of the famous Dr. Sydenham of laudanum fame and that later by the rescue of Alexander Selkirk in February, 1709 he inspired Danial De Foe's classic piece of literature '*Robinson Crusoe.*' In 1718, Dover retired as a wealthy buccaneer and at the age of 40 became a successful medical practitioner. He used mercury in such large doses that he became known as a "quicksilver" doctor. He usually gave Dover's powder in 60-grain doses instead of five and is claimed to have given as high as 100 grains, the equivalant of 10 grains of opium and 10 grains of Ipecac.

Dover's powder was made famous by Joshua Ward, the renowned quack, a protege of King George II of England and a friend of such well-known personages as Chesterfield, Gibbon, Fielding, Reynolds, and Walpole.

The history of the chemistry of opium is not as old as that of opium poppy or opium itself. Raw opium is a complicated chemical mixture containing fats, acid, mucilage, proteins, and about 25 or more alkaloids. Significant research of three European pharmacists, Derosine, Seguin, and Serturner in the first decade of the last century led to the discovery of the first known alkaloid, morphine. Derosne, a French chemist, produced in 1803 a crystallizable salt from opium and thought that he had discovered the active basic principle of opium. His product, called Sel Narcotigue de Derosne, was later found not to have narcotic properties. Another French chemist, Seguin, an assistant to Fourcroy, isolated the active principle in 1804. He was on the trail of morphine. Apparently, he did not realize its significance because his paper was not published for ten years.[1]

Meanwhile in 1806, Friedrich Wilhelm Sertürner, a German pharmacists' assistant, reported the discovery of meconic acid, a new organic acid in opium. He experimented with this acid on animals, and later announced the discovery of the narcotic principle of opium. He named it Morphinum from Morpheus, the god of dreams and servant of Somnus, the god of sleep. The word Morpheus means "the fashioner or molder," arising from the same etymological stem as "morphology."

At an early date it had been recognized that the poppy plant had the power of calling up strange shapes or dreams. What part the drug had in the creation of shapes and forms is evident in the now famous works of De Quincey,[41] Poe, and Coleridge.[42]

Morphine constitutes about 10 percent of the alkaloids (of opium only). Stimulated by this discovery, many vegetable drugs were investigated and discoveries such as quinine, strychnine, emetine, nicotine, and picrotoxin were announced. Pierre Jean Robiquet (1780-1840), a French physician, a pupil of Fourcroy, and still later of Vauquelin, isolated codine or methyl morphine, present 0.1 to 3 percent in opium. Heroin or diacetylemorphine was prepared at a later date by heating morphine with acetic anhydride.

In the thirteenth and fourteenth centuries leading European surgeons were recommending some of the soporifics and narcotics for surgical operations. In America, John Leigh in 1875 won the Harveian prize for his essay on experimental enquiry into the properties of opium.

This tied in well with the American interest in anesthesia. Abraham Jacobi has stated, "The greatest gift America has given to the World is not the realization of the republican government-ancient culture exhibited it before and allowed it to perish by political short-sighted lust of conquest and the undemocratic jealousy–it is anaesthesia."

REFERENCES

1. Scott, J. M. The White Poppy–A History of Opium. Funk and Wagnalls, New York, 1969.

2. Kritikos, P. G. and Papadaki, S. P. The History of Poppy and of Opium and their Expansion in Antiquity in the Eastern Mediterranean Area. Bull Narcotics, Volume XIX, No. 3, 1967.

3. Neligan, A. R. The Opium Question. Bale and Curnew, 1927, London.

4. Anslinger, H. J. and Tompkins, W. F. The Traffic in Narcotics. New York, 1953.

5. Dougherty, R. P. Cited in Terry and Pallens, The "Opium Problem." New Jersey, 1970.

6. Thompson, R. C. Assyrian Herbal. London, 1924, pp. 46, 261-269. Assyrian Herbal is a translation of an Assyrian Medical Tablet now placed in the British Museum.

7. Fields, A. The Merk Report. 1949, pp. 58, 2.

8. Galen, Claudii Galeni. Medicorum Graecorum Opera Quae Extant. Leipzig, 1826.

9. Bryan, C. P. "Ebers Papyrus." Translated from German version. Geoffrey Bles, London, 1930.

10. Dai, Bingham. Opium Addiction in Chicago. Century Press Shanghai, 1937 (PhD Thesis).

11. Woenig, Franz. Die Pflanzen des Alten Aegypten. Leipzig, 1886, p. 225. Cited by Kritikos, P.G. and Papadaki, S.P.

12. Schweinfurth. Cited by Saber Gabra in "Papaver Species and Opium through the Ages." Bulletin de l' Institut de Egypte, 1956, p. 40 and H. R. Hall, Journal of Egyptian Archaeology, 1928, 14 pp. 203-205.

13. Gabra, S. Papaver Species and Opium through the Ages. Bulletin de l' Institut de Egypt, 1956.

14. Emboden, W. Narcotic Plants. Studio Vista, London, 1979.

15. Robinson, Victor. The Story of Medicine. New Home Library, New York, 1943.

16. Giles, H. A. Historic China and Other Sketches. T. De La Rue, London, 1882.

17. Eldein, Dr. Watts' 1908 Commercial Products of India. Reprint. Today and Tomorrow, New Delhi, 1966.

18. Husain, Akhtar and Sharma, J. R. The Opium Poppy. Central Inst. of Medicinal and Aromatic Plants, Lucknow, 1983.

19. Hesiod, Theogony II, pp. 535-537. Cited by Kritikos, P. G. and Papadaki, S. P.

20. La Wall, C. H. The Curious Lore of Drugs and Medicine. Garden City Publication Company, Garden City, 1927.

21. Homer. Iliad IX, pp. 306-307. Cited by Kritikos, P. G. and Papadaki, S. P.

22. Homer. Odyssey X, pp. 220-232. Cited by Kritikos, P. G. and Papadaki, S. P.

23. Hippocrates. 460-359 B.C. Oeuvres Completes d'Hippocrates Littre, Paris, 1840-1849. Cited by Kritikos, P. G. and Papadaki, S. P.

24. Theophrastus. 372-287 B.C. History of Plants IX 15-1. Cited by Kritikos, P. G. and Papadaki, S. P. "The history of poppy and of opium and their expansion in antiquity in the Eastern Mediterranean area. Bull Narcotics. Vol. XIX, No. 3, 1967.

25. Gunther, Robert T. The Greek Herbal of Dioscrides. Oxford University Press, London, 1934.

26. Aristotle. Physica Minora 456B, 30. Historia Animalium 1, 6276, 18. Cited by Kritikos, P. G. and Papadaki, S. P.

27. Tschirch, Alex. Handbuch der Pharmakognosi. 111/1 Leipzig, 1923, p. 647. Cited by Kritikos, P. G. and Papadaki, S. P.

28. Pliny, The Elder. NHXX77. Cited by Kritikos, P. G. and Papadaki, S. P.

29. Lewin, L. Narcotics and Stimulating Drugs. Phantastica, 1931, pp. 36-37.

30. Fields, Albert. Med. Rec., 1948, pp. 161, 232.

31. Fields, Albert. Military Surgeon, 1948, pp. 103, 249.

32. Marti Ernest, Prof. Cited by A. Tschirch, Handbuch des Pharmakognosie I/III, 1933, p. 1208.

33. Rindley. Cited by A. Tschirch. Handbuch des Pharmakognosie I/III, 1933, p. 1209.

34. Walker, Winifred. The Plants of the Bible. London, 1959, p. 88.

35. Dwarkanath, C. Use of Opium and Cannabis in the Traditional Systems of Medicine in India. Bull. Narcotics, 1965, XVII, Vol. 15.

36. Sharma, P. V. Drugs and the Landmarks of the History of Indian Medicine. Jour. Res. Indian Med., 1973, 8(4), p. 86.

37. Asthana, S. N. The Cultivation of Opium Poppy in India. Bull. Narcotics 1954, VI 3-4.
38. Chopra, R. N. and Chopra, I. C. Quasi-medical Use of Opium in India and Its Effects. Bull. Narcotics, 1955, VII (3, 4), 1-22.
39. Sigerest, A. E. Paracelsus in the Light of Four Hundred Years. March of Medicine, NY Academy of Medicine, NY, 1941.
40. Terry, C. E. and Pellens, M. The Opium Problem. Haddon Craftsmen, Camden, NJ, 1928.
41. Burke, Thomas. The Ecstasies of Thomas De Quincey. Harrap, London, 1928.
42. Charpentier, John. Coleridge the Sublime Somnambulist. Constable, London, 1929.
43. Veselovskaya, M. A. The Poppy, Amerind Publishing Co. Pvt. Ltd., New Delhi, Translated from Russian, 1976.
44. Bisset, N. G. Plants as source of Isoquinoline alkaloids. In Chemistry and Biology of Isoquinoline alkaloids. Ed. by J. D. Phillipson & M. F. Robert, Springer-Verlag (1985).
45. Schultze, Motel J. (1979) Die Urgeschichtlichen Reste des Schlafmohns (Papaver somniferum L) und die Entstehung der Art. Kulturpflanze 27:207-215.
46. Krikorian, A. D. (1975) were the opium poppy and opium known in the ancient near East J. Hist Biol 8:95-114.
47. Merrillees, R. S. (1979) opium again in antiquity. Levant 11:167-171.
48. Merrillees, R. S. (1962) opium trade in Bronze Age Levant. Antiquity 36:287-292.
49. Merrillees, R. S. (1968) The Cypriot Bronze Age pottery found in Egypt. Stud Mediter Archaeol 18:154-157, 176, 179, 196, Pl, XXXV 1 & 2, XXXVI, 1.
50. Muzio, 1 (1925) Su di un olio medicate della Tomba di cha-Atti. soc Lingustica Sci Lett 249-253.
51. Schiaparelli E., (1927) La Tomba intatta dell' architetto Cha nella necropoli di Tebe in Relazione sui lavori della missone archeologica Italiana in Egitto (anni 1903-1920) Vol. II. R. Mus Antichita, Torino, pp. 154-158.
52. Vesselovskaya M. A. (1975) Poppy (Variability, classification, evolution) Tr Prikl Bot Genet Sel. 55(i): 175-223.
53. Hammer K., Fritsch R. (1977) Zur Frage nach der Ursprungsart des Kulturmohns (Papaver somnferum). Kulturpflanze 25:113-124.
54. Hammer K. (1981) Problems of *Papaver somnferum* classification and some remarks on recently collected European poppy land races. Kulturpflanze 29:287-297.
55. Danert. S. (1958) Zur systematic Von Papaver somniferum L. Kulturpflanze 6:61-88.

Chapter 2

Botanical Studies

The genus *Papaver* comprises about 110 species with many varieties and forms mostly native to Central, South Europe, temperate and sub-tropical regions of the Old World. Many species are annuals but quite a number are perennial herbs.[1] Fedde[2] recognized 99 species of the genus *Papaver* and divided them into nine sections. *Papaver somniferum* belongs to the section 'Mecones.'

Systematic position of the genus *Papaver* has been recorded by Trease & Evans[3] as below.

Papaverales: Papaveraceae, Capparaceae, Cruciferae, Resedaceae—an order of six families, or seven if the Fumariaceae is separated from the Papaveraceae. The Papaveraceae belongs to the suborder Papaverineae and the Capparaceae and Cruciferae to the Capparineae. Some workers regard the Papaveraceae as related to the Ranunculales and the Capparineae as derived from the Cistales. Chemical support for this view is that alkaloids of the Papaveraceae are related to those of the Ranunculaceae, and that thiogluconates are absent from the Papaveraceae but present in the other two families.

The Papaveraceae (including Fumariaceae) is a family of 42 genera and about 650 species. The plants are usually herbs with solitary, showy flowers of the floral formula K 2 - 3, C 2 + 2 or 2 + 4, A ∞, G (2 $-\infty$). The fruit is generally a capsule, with numerous seeds, each containing a small embryo in an oily endosperm. Genera include *Platystemon* (about 60 spp.), *Romneya* (2 spp.), *Eschscholtzia* (10 spp.), *Sanguinaria* (1 sp.), *Chelidonium* (1 sp.), *Bocconia* (10 spp.), *Glaucium* (25 spp.), *Meconopsis* (43 spp.), *Argemone* (10 spp.), and *Papaver* (100 spp.). The group containing *Fumaria* (55 spp.) also contains *Corydalis* (320 spp.) and *Dicentra*

(20 spp.). All members contain latex tissue. Sometimes the latex is in vessels which accompany the vascular system (e.g., in *Papaver*); sometimes in latex sacs (e.g., *Sanguinaria*). As will be seen from the research references, the family is rich, both in number and variety, in alkaloids. Some, such as the opium alkaloids, are of great medical and economic importance.

Stermitz[14] observed that a standard botanical reference for the genus *Papaver* is that of Fedde[19] who recognized 99 species which were divided into nine sections–Argemonorhoeades, Carinatae, Horrida, Mecones, Miltantha, Orthorhoeades, Oxytona, Pilosa, and Scapiflora.

The genus *Papaver* was originally divided[16] into five sections– Lasiotrachyphylla, Oxytona, Miltantha, Rhoeades, and Mecone. El- kon[17] introduced the section Horrida, and Prantl and Kunding[18] included the section Pilosa. In the section Rhoeades those species were classified which Fedde[19] had placed in the Orthorhoeades and Argemonorhoeades, and excluded the section Carinatae. The divi- sion of the genus *Papaver* into nine sections is still used by many authors. Gunther[20] divided the section Pilosa and to the genus *Pa- paver* he added the genus Roemeria as a special eleventh section. Later the section Mecones was divided into the section *Papaver* and the section Glauca.[21,22] (See Table 2.1.)

According to the new classification[21,22] the section *Papaver* (Syn-Mecones) includes the species *P. somniferum, P. setigerum* DC, *P. glaucum* Boiss et Hausskn; *P. gracile* Auch and *P. decaisnei* Hochst & Steud. Systematic investigations have shown that the original section *Papaver* (syn. Mecones Bernh) is rather heteroge- nous in its chemical composition. The presence of morphinane alkaloids, thebaine, codeine and morphine together with the se- cophthalideisoquinoline alkaloids narceine, nornarceine, and nar- ceine imide and the phthalideisoquinoline alkaloids narcotine, and narcotoline is characteristic of *P. somniferum* L and *P. setigerum* DC. In contract rhoeadines and papaverrubines predominate in *P. glaucum, P. gracile* and *P. decaisnei* Hochst et steud. Traces of rhoeadines are also found in *P. somniferum,* and *P. setigerum* whereas morphinanes have not been found in *P. glaucum* and *P. gracile.* Minor alkaloids of morphine, rhoeadine, and coptisine

TABLE 2.1. Retrospective historical survey of the sections of the genus *Papaver*

Bernhardi (16)	Elkan (17)	Prantl (18)	Fedde (19)	Günther (20)	Present State
Lasiotrachyphylla Bemh.	*Scapiflora* Reich	*Lasiotrachyphylla* Bemh.	*Scapiflora* Reich	*Lasiotrachyphylla* Bemh.	*Meconella* Spach.
Oxtona Bemh.	*Macrantha* Elk.	*Macrantha* Elk.	*Macrantha* Elk.	*Oxytona* Bemh.	*Macrantha* Elk.
		Pilosa Prantl	*Pilosa* Prantl	*Pilosa* Prantl	*Pilosa* Prantl
Miltantha Bemh.	*Pyramistigmata* Elk.	*Miltantha* Elk.	*Miltantha* Bemh.	*Miltantha* Bemh.	*Miltantha* Bemh.
Rhoeades Bemh.	*Rhoeades* Bemh.	*Rhoeades* Bemh.	*Argemonorhoeades* Fedde	*Argemonidium* Spach.	*Argemonidium* Spach.
			Carinatae Fedde	*Carinatae* Fedde	*Carinatae* Fedde
			Orthorhoeades Fedde	*Rhoeades* Bemh.	*Rhoeadium* Spach.
Mecones Bemh.	*Mecones* Bemh.	*Mecones* Bemh.	*Mecones* Bemh.	*Papaver*	*Papaver*
					Glauca J. Nov. et V. Prein.
	Horrida Elk.	*Horrida* Elk.	*Horrida* Elk.	*Horrida* Elk.	*Horrida* Elk.
				Pseudo-pilosa M. Pop.	
				Roemeria (Medic.) Günther	*Roemeria* (Medic.) Günther

21

are found in poppyheads of *P. decaisnei* where the major alkaloid is papaverine.[15]

GENERAL DESCRIPTION OF THE PLANT

Since opium poppy, *Papaver somniferum*, is a cultivated plant, it shows considerable variations in the color of flower, seeds, and the shape of the capsule. However, the general appearance of the plant does not vary so much and may be described as below.

Annual herbs have thick tapering roots and stems reaching a height of 3-4 ft., erect, cylindrical, solid, and quite smooth. The leaves are large, numerous, alternate, sessile, and clasping the stem by a cordate base. The buds are ovate and drooping but the flowers are erect, solitary, and large with floral formula K2,C2 + 2,A∞, G(2 −∞). Two sepals which are green, broad, and quite smooth disarticulate and are pushed away as the flower opens. The four large petals are decussate, the outer two are wider and much overlapping slightly narrower inner ones. They are concave, undulated with numerous closely placed veins radiating from the stiff thick wedge-shaped base. They are pure snow white and glossy.

The stamens are numerous, hypogynous, inserted in two to three rows on the undersurface of the dilated thalamus. Filaments are long, flat, and ribbon shaped, slightly dilated at the top. Anthers are linear and attached by a narrow base to the filaments. They are cream colored, wavy, and are twisted after dehiscence.

The ovary is large and globular but contracted below into a neck (gynophore) which again dilates to form the receptacle and this also narrows off below into the pedicel. The latter is quite smooth and green in color. The ovary is unilocular and contains large spongy parietal placenta which bulges out nearly to the center. The placenta are almost always equal to the number of stigmatic rays and bear numerous ovules over all parts of their surface. The stigma is sessile, peltate, and spreading over the top of the ovary with 8 to 13 short obtuse oblong rays.

The fruit, a capsule, is usually more or less globular, supported on a neck (as in ovary) and crowned by persistent stigma. The pericarp is hard, smooth, dry, and brittle, and brownish-yellow when ripe. It has one central cavity with dry papery placental plates reaching about halfway to the center.

Capsules of some varieties are dehiscent and of others indehiscent. The seeds are small but numerous, reniform, and yellowish-white in color. The testa has a reticulated network and the embryo is slightly curved within the oily endosperm.

When the bud unfolds and the whole flower is open during the day, the sepals tend to drop first and the petals fall off after 24 to 48 hours. The stamens may persist for a short period after the petals drop but they soon dry up. The ovary, after fertilization, and after the petals have dropped, develops in size and within two weeks assumes considerable dimensions when it can be considered fit for lancing.

Five species occur in Great Britian. *P. rhoeas* is the common scarlet poppy found in fields. Cultivated forms of this species with exquisite shades of color and without any blotch at the base of the petals are known as Shirley poppies.

Asthana[4] reported the presence of *P. croceum* with flowers of orange color as the only species growing wild in alpine regions of India at altitudes of 12000-18000 ft. Among the other species reported by him are *P. hybridum* and *P. pavonium* growing with prickly capsules as annuals in the plains of India. The former has red brick flowers and the latter has larger salmon or orange flowers with a dark center. Another species *P. dubium* var. *glabrum* has orange or salmon flowers with a dark spot near the base of each petal. *P. marcostonum, P. turbinatum,* and *P. rhoeas* are common in the wheat fields of Kashmir. The following are the three main varieties of *P. somniferum.*

1. *P. somniferum* var. *nigrum* DC

A form of the opium poppy with purple-red flowers, roundish oblong capsule, opening by pores under the stigma with seeds of dull greyish-black color.

2. *P. somniferum* var. *album* DC

Also a form of opium poppy with white flowers, roundish ovate capsules not opening by pores under the stigma, seeds white.

3. *P. somniferum* var. *abnormale*

A variety not infrequent in neglected poppy fields. Flowers small, streaked with dull green and red, the petals much crumpled and never expanding fully; capsules roundish oblong, opening by pores under the stigma.

These varieties in their natural state are considered to be poor drug yielding varieties unless improved by cultivation.

DISTRIBUTION

Papaver somniferum is not found growing wild and is mostly domesticated. It has been successfully grown in such diverse areas as Europe, northeast Africa, Australia, Japan, South America, and North America. But due to economic and other considerations it is legitimately cultivated in India, China, Egypt, France, Holland, Hungary, Greece, Spain, Portugal, Italy, Turkey, Australia, Russia, and the former Yugoslavia.[5] Japan and Bulgaria also grow opium poppy on a limited scale. It is also cultivated in many other parts of the world, such as Thailand, Burma, Laos, Afganistan, Pakistan, Iran, Austria, Czechoslovakia, Germany, Poland, and Romania.[6]

Quite a number of botanists have collected *Papaver* species from different localities in India. A catalogue of herbarium sheets of Indian species indicates that the indigenous species are mainly distributed in northwestern India and some are sparingly found in eastern parts of India. The catalogue received through the courtesy of curator Indian Botanical Garden Calcutta is reproduced in Table 2.2.

All plants of the family Papaveraceae are reported to be rich in alkaloids.[8] Six species occur in India but *P. somniferum* is cultivated as the chief source of commercial opium.

A number of species of *Papaver* are grown as ornamental plants for their beautiful flowers ranging in color from white to almost black through various shades of yellow, pink, orange, scarlet, and crimson. The species most commonly grown are *P. nudicaule* (Iceland poppy), *P. orientale* (oriental poppy), *P. rhoeas* (corn poppy), and *P. somniferum* (opium poppy). Many varieties, strains, and hybrids of poppies with single or double flowers and some with fringed petals have been raised. Poppies can be cultivated in open situations with rich loamy soil. Seeds are sown broadcast and seedlings are thinned out. Flowers appear after one and a half to two months. *P. oriental* can be propagated by root cuttings. A brief description of these species is given beginning on page 28.[1]

TABLE 2.2. A catalog of the herbarium sheets of Indian species of *Papaver*

Sr. No.	Name	Locality	Date of Collection	Collector's Name	Field No.
1	2	3	4	5	6
1	*Papaver* nudicaule L.	Ladakh	–	T. Thomson	–
2	*Papaver* nudicaule L. var. rubro-aurantiaca Fischer	–	–	Surg. Capt. Alcock	17686
3	*Papaver* nudicaule L.	Khardungle Ladakh	19.9.1955	B.K. Abrol	4490
4	*Papaver* nudicaule L. var. rubro-aurantiaca Fischer	Tomtek Jilqa	11.7.1913	Lieut. Kenneth Mason R. E.	18
5	*Papaver* nudicaule L.	Nithaivally Gilgit Kashmir	4.8.1892	J. F. Duthie	–
6	*Papaver* nudicaule L.	North Sonamarg, Kashmir	15.8.1913	Capt. F. E. Koebel	126
7	*Papaver* nudicaule L.	Ladakh	–	B. D. Keyde	–
8	*Papaver* nudicaule L.	–	10.6.1880	J. E. T. Aitchison	104
9	*Papaver* nudicaule L.	Pangi, N. W. Himalaya	1879	Rev. A. W. Heyde	–
10	*Papaver* nudicaule L.	Bujila, Kashmir	1877	C. B. Clarke	29929
11	*Papaver* nudicaule L.	Sangum, Valley Kashmir	27.7.1893	J. F. Duthie	–
12	*Papaver* nudicaule L.	Kargeh Valley, Kashmir	30.8.1893	J. F. Duthie	–
13	*Papaver* nudicaule L.	West Tibet	1864	Falconer	112

TABLE 2.2 (continued)

Sr. No.	Name	Locality	Date of Collection	Collector's Name	Field No.
1	2	3	4	5	6
14	*Papaver* nudicaule L. var. grand-iflorum N. var.	Chumbi	1877	G. King	4531
15	*Papaver* nudicaule L.	Tibet	1882	Dr. King's Collection	36
16	*Papaver* nudicaule L. grand-iflorum N. var.	Tibet	1882	Dr. King's Collection	36
17	*Papaver* hydridum L.	Punjab	–	–	–
18	*Papaver* hybridum L.	Punjab	9.1892	J. E. T. Aitchison	63
19	*Papaver* hybridum L.	Hazara	–	Stewart	–
20	*Papaver* macrostomum Boiss.	–	10.7.1876	Wallich	8120
21	*Papaver* dubium L. = *Papaver* macrostomum Boiss.	Ramoo Kashmir.	–	C. B. Clarke	28543
22	*Papaver* macrostomum Boiss.	Kashmir	3.7.1902	I. R. Dreemond	13999
23	*Papaver* macrostomum Boiss.	Lahore, N.W. Himalaya	11.3.1934	Dr. Stewart	2580
24	*Papaver* macrostomum Boiss.	Srinagar, Kashmir	8.5.1892	J. F. Duthie	10838
25	*Papaver* roheas.L.	Chambak	8.6.1864	Dr. Brandis	4336
26	*Papaver* hookeri Baker.	Prutapotum	22.11.1887	J. R. Drummond	6293
27	*Papaver* roheas.L.	Darjeeling	–	Kurz.	–
28	*Papaver* roheas.L.	Tibet	1882	Dr. King's Collection	36

Sr. No.	Name	Locality	Date of Collection	Collector's Name	Field No.
1	2	3	4	5	6
29	*Papaver* somniferum L.	–	–	–	1412
30	*Papaver* roheas var. Latifolia Ham. = P. Hookeri Bak.	Cult. Hort. Bot. Calcutta	–	Wallich	1856
31	*Papaver* roheas var. latifolia Ham. = P. Hookeri Bak.	Missoorie	–	G. King	–
32	*Papaver* dubium L. var. glabrum Koch = P. laevigotum M. Biel	Punjab Himalaya	May, 1884	Dr. D. D. Cunningham	–
33	*Papaver* dubium L. var. glabrum Koch = P. laevigotum M. Biel	Kumaon Jeolikote	5.2.1913	N. Gill	547
34	*Papaver* dubium L. var glabrum Koch = P. laevigotum N. Biel	Tehri Garhwal N. W. Himalaya	18.5.1897	J. F. Duthie	19825
35	*Papaver* dubium L.	Cornfield near Ghaniss, Simla.	April, 1886	E. R. Johnson	–
36	*Papaver* dubium L. var. glabrum Koch.	Above, Sillagat Chamba State, N. W. Himalaya	25.5.1896	J. H. Lace	1397
37	*Papaver* dubium L. var. glabrum Koch.	Sannsir, N. W. Himalaya	12.5.1893	J. F. Duthie	12933
38	*Papaver* dubium L. var. glabrum Koch.	Gurhwal	1864	Falconer	113

TABLE 2.2 (continued)

Sr. No.	Name	Locality	Date of Collection	Collector's Name	Field No.
1	2	3	4	5	6
39	*Papaver* dubium L. var. glabrum Koch.	Baruni, Simla	17.5.1876	N. Gamble	4233c
40	*Papaver Somniferum* L.	Saraon Patti Maja Road	16.3.1896	D. Prain	–
41	*Papaver Somniferum* L.	Bhandwa, Maja Road	1.3.1901	Mrs. A. S. Bell	48
42	*Papaver Somniferum* L.	Bhandwa, Maja Road	16.3.1896	D. Prain	–
43	*Papaver Somniferum* L.	Balek to Lokpur	9.3.1912	R. E. P.	36997
44	*Papaver Somniferum* L.	Munipur, Assam	5.1882	George Watt	7208
45	*Papaver Somniferum* L.	Choongthang Sikkim, Himalaya	5.1885	King's Collection	–
46	*Papaver* croceum Ledeb.	Kashmir	10.1902	J. R. Drummond.	14824

Courtesy of Indian Botanical Garden, Calcutta.

The table indicates that *P. somniferum* does not occur in the wild state of growth. The collection under item No. 29 in the table is not clear and may be from cultivated plants or as an escape.

1. *Papaver argemone* L.

An annual herb indigenous to Europe and the Mediterranean region and commonly grown in gardens in India. Flowers pale scarlet; capsules oblong elliptical. Flowering plant contains 0.15 percent of alkaloids, including rhoeadine, protopine, and an unidentified alkaloid.

2. *P. dubium* L.

An annual herb found in the West Himalayas from Kashmir to Garhwal. Flowers red with a dark spot at the base of the petal; fruits obovoid capsule, smooth, seeds numerous, small, kidney shaped.

The latex from immature capsules contains two alkaloids, viz. aporeine ($C_{18}H_{16}O_2N$., m.P. 88-89°), aporeidine (m.p. 176-78°), and meconic acid. According to recent investigations, the plant contains rhoeagenine as the principle alkaloid, besides rhoeadine, protopine, and two unidentified alkaloids (m.p. 159-61° and 243-34° decomp. respectively).[1]

The table indicates that *P. somniferum* does not occur in the wild state of growth. The collection under item No. 29 in the table is not clear and may be from cultivated plants or as an escape.

3. *P. nudicaule* L. (Iceland Poppy)

A perennial hairy herb found in Gulmarg (Kashmir) at 9000-1200 feet commonly treated as an annual garden plant.

Flowers may be white, yellow, orange, or red in color. The capsules contain cyanogenetic glucoside and an emulsion-like enzyme, plants with yellow flowers contain more cyanogenetic glucosides than those with red or white flowers. Fresh leaves yield 3.1 to 5.1 mg./1 ∞ g of hydrocyanic acid.[1]

4. *P. orientale* L. (Orientale poppy)

An erect perennial herb indigenous from the Mediterranean region to Iran and grown in gardens of India. Flowers scarlet with a black violacious spot at the base of petal; capsule glaucous, subglobose; seed orbicular-reniform, brown broadly striate. The plant contains 0.16 percent of alkaloids which include thebaine ($C_{19}H_{21}O_3N$., m.p. 193°), isothebaine (m.p. 203-4°), protopine ($C_{20}H_{19}O_5N$., m.p. 207°), glaucidine (m.p. 209-10°), and oripavine ($C_{18}H_{21}O_3N$., m.p. 200-1°).

Thebaine is reported to be the predominant alkaloid during active growth of plant but at maturity the plant contains mostly isothebaine. Presence of potassium nitrate also has been reported.[1]

5. *P. rhoeas* L. (Corn poppy)

An erect, branched, very variable annual found in the fields of Kashmir. Flowers scarlet, with dark eye, capsules subglobose, smooth, seeds dark brown.

P. rhoeas is commonly found growing as a weed in gardens and one of its strains known as Shirley poppy is the most popular of ornamental poppies. It has flowers with exquisite shades and without any blotch at the base of the petals. An alkaloid rhoeadine ($C_{21}H_{21}O_6N.$, m.p. 256-575° in vacuo) is present in all parts of the plant including the roots. The capsule also contains morphine, thebaine, narcotine, and meconic acid. Protopine and coptisine, besides a number of other phenolic and non-phenolic crystalline and amorphous bases, have been reported in the roots and aerial parts. In addition rhoeagenine (m.p. 240-43°) and a number of other uncharacterized alkaloids were also isolated.[1]

6. *P. somniferum* L. (Opium poppy, White poppy)–Vernacular names:[9]

Arabic: Abunom, Afiun, Bizrulkhashkhash, Khashkhashullaiza, Qishrulkhashkhash-; *Bengal:* Pasto, Post-; *Bombay:* Aphim, Appo, Khaskhas, Post-; *Burma:* Bhain, Bhainzi-; *Canarese:* Afim, Biligasgase, Gasagase, Khasakhasi-; *Catalan:* Cascall-; *Chinese:* Ying Tzu Su-; *Danish:* Valmiue-; *Decan:* Afim, Khashkhash-; *Dutch:* Heul, Slaapkruid; *English:* Bale-wort, Carnation Poppy, Joan Silverpin, Opium Poppy, Peony Poppy, White Garden Poppy, White Poppy-; *French:* Pavot, Pavot blanc, Pavot des jardins, Pavot a opium, Pavot somnifere-; *German:* Mahn, Saatmohn, Schlafmohn-; *Greek:* Agria, Mikon hymeros-; *Gujarat:* Aphina, Khuskhus, Posta-; *Hindi:* Afin, Afyun, Kashkash, Pest, Post, Postekebij-; *Indo China:* A phien, A phu dung, Co tu tue-; *Italian:* Papavero, Papavero domestico-; *Kachhi:* Doda, Post-; *Kumaon:* Posht-; *Malaya:* Bungapion, Yin soo hock-; *Malayalam:* Afiun, Kashakasha-; *Malta:* Opium Poppy, Poppy, Papavero, Pianta da oppio, Papavru, Xahxieh-; *Marathi:* Aphu, Khuskhus, Posta-; *Nepal:* Aphim-; *Oudh:* Posta-; *Persian:* Afiun, Khashkhash, Khashkhashsufaid, Koknar, Postekoknar, Tukhmekoknar-; *Portuguese:* Dormideira, Papoula branca-; *Punjab:* Afim, Doda, Khashkhash, Khishkhash, Post-; *Romanian:* Mac, Mac somnisor, Somnisor-; *Russia:* Mak Snotvornyi-; *Sanskrit:* Ahifen, Chosa, Khasa, Khakasa, Ullasata-; *Sinhalese:* Abin-; *Spanish:* Adormidera Dormidera-; *Swedish:* Vallmo-; *Tamil:* Abini, Gashagasha, Kasakasa, Postaka-; *Telugu:* Abhini, Gasagasala, Gasalu, Kasakasa-; *Turkish:* Hashish-; *Urdu:* Khashkhashsufaid.

These names stand for plants, capsules, and seeds.

There are numerous varieties of *P. somniferum*, but only two are under cultivation.

P. somniferum var. *album* DC with ovate-globose capsules devoid of apertures is cultivated in India, parts of Iran, and the former Yugoslavia. *P. somniferum* var. *glabrum* DC with red, purple, or variegated flowers and almost spherical capsules dehiscing through opening below the stigmatic lobes is cultivated in Asia Minor, Egypt, Turkey, and parts of Iran. *P. somniferum* var. *nigrum* DC with open capsules is particularly cultivated for seeds in Europe.

There are several forms of var. *album* under cultivation in India. One with white flowers and white seeds is grown in Uttar Pradesh, the form with red or purple flowers is grown in Madhya Pradesh and Rajasthan (Malwa), and a partly colored form is found in the Himalayas.

Various races of opium poppy are grown in India but no comprehensive classification has been so far evolved.[4] They vary in size and shape of plants, leaves, petals, capsules, and in drug-yielding capacity. Sometimes there is a marked difference in the color and quality of latex obtained, its morphine content, and in the ratio of the principle alkaloids present in it. Since cross-pollination is common, a crop of opium poppy contains a wide range of forms composed of hybrids.

Various races of Indian opium poppy may be separated into well-defined groups based on the color and texture of capsules.

(a) Subza-dheri race.

Capsules opaque, green in color with deeper or paler shades.

(b) Sufaid-dheri race.

Capsules glaucous, more or less densely coated with opaque white powder.

These races are known in India by their local names.

The *"subza-dheri"* is an early variety whereas the sufaid-dheri race may be used for late or early sowing. *Subza-dheri* yields about half of the normal flow of latex during hot sunny days. The white coating powder on the *Sufaid-dheri* helps to resist evaporation during high temperature and dry atmosphere.

The races of *P. somniferum* grown in Uttar Pradesh are:

1. *Telyleah* variety or *Telia, Haraina, Hariala,* or *Herera.* It has an oblong ovate capsule with pale green color without powdery coating. It is quite productive as an early crop.

2. *Sufaid-danthi* or Katha Bhabutia. It is very popular for late sowing but less productive than the Teyleah variety.

3. *Kutila, Katila,* or *Kotila,* or *Chansura, Ghanghabaga, Chirrah, Bhagbhora.* These races can withstand hailstorms and high winds and is resistant to blight, suitable for sandy loam. Capsules are oblong-ovate and glaucous. The latex at the time of lancing is red.

4. *Choura Kutila.* A good race suitable for clayey loam, requiring more moisture than other varieties.

5. *Kaladanthi, Karria, Damia, Kalidanthi,* or *Kalidandi.* It is a very popular race with farmers. It is well marked by the peculiar bluish-black color acquired by the flower stalk soon after the fall of the flower. It has a short life cycle but gives more opium and thus better returns to the farmer. The capsule is oblong-obovate and glauceous. The latex is slightly reddish at the time of lancing.

6. *Subza kaladanthi* or *Haraina kaledanthi.* Not a popular race. The capsule is olive-green in color, sensitive to high temperature and excess moisture.

7. *Kaledanthi Baunia.* It resembles *kaladanthi* (5), however, the plants are smaller but produce more opium than the *kaladanthi* variety.

8. *Monoria.* It yields a fairly good quantity of opium but requires a well-manured soil or clay-loam. It has large-roundish, ovate, and glaucous capsules.

9. *Dheri-Danthi.* An off-shoot of Sufaid-dheri. Resistant to blight disease but does not produce desired quantity of opium which, however, is rich in morphine content.

10. Variegated poppy. A form of *sufaid-dheri* resistant to blight and pests which may be due to more highly oxygenized state of tissues than that of normal forms.

11. *Sufaid-danthi monoria.* This is a hybrid, quite robust and produces large uniform capsules of roundish oblong shape.

12. *Monoria Teyleash.* This also is a promising hybrid. It resembles its female parent *monoria* "in general appearance

and form of capsule but its texture is that of the "teyleash" type and the color is deep opaque green.

13. *Sandbha* or *Dhadhus* or *Bhabhua.* The plants are comparatively taller with big roundish oblong capsules. The yield of opium is comparatively very little.
14. *Sahbania.* This variety is grown in eastern districts. It has long leaves with the two basal lobes falling down the leaf stalks. Capsules oblong, ovate with rough surface. It does not yield much opium.

The races of opium poppy cultivated in Rajasthna and Madhya Pradesh differ from that of Uttar Pradesh in some characteristics. In general, the poppy in these regions has a straight stalk, simpler stem, sharply toothed leaves of much thinner texture, usually red- or purple-colored flowers with fringed petals and large-sized oblong or ovate-oblong capsules with broad stigmatic rays at the top. The following are the main races grown in this region.

1. *Bhatphoria or Dhaturi.*
   Plant Height 3'6," capsule 3" × 2 1/2" roundish elongated, light green in color. Opium yield rather poor.
2. *Galania.*
   Height 4' or less. Flowers white with pink or dark pink border. Capsule size 3" × 2". Oblong and flattened a little on top, color dark green. Yields more opium and less seed than *Dhaturia.* Color of opium light brown.
3. *Hybrid of the above two races.*
   Petals red and white, often white, but mixed also.
   Average height of plant, 4 feet.
4. *Ramazatak.*
   Flowers white and red-white in color, capsule small and elongated, slightly flattened at the top, size 3" × 2". Opium yeild more and seed less in comparison to *Dhaturia.* Color of opium, dark brown turning black.
5. *Telia.*
   Flowers white. Petals 2 1/2" long non-furcated. Capsules elongated, light green and shining. Opium of dark shade. Seeds white. Produces more seed than *Ghotia* or *Chaglia.*

6. *Chaglia.*
Petals 2" × 1/2", red or pink with white dots at the bottom. Major portion of the petal is colored. Shade of opium lighter than No. 1 and 2. Average yield of opium highest of all. Seeds white.
7. *Kasturi or Tejani.*
Scarlet red flowers. Very poor yield of opium, seeds red.
8. *Ghotia.*
Petals 2 1/4", white or with pink border. Capsules round, dark green. Color of opium as in Telia. Produces less seed than *Telia* but more than *Chaglia.*

The cultivars of opium poppy have been studied from different altitudinal zones of the northern hemisphere and have been classified on this basis by Veselovskaya.[5] The crop is raised at different times of year suiting the different climatic zones. In India opium poppy is a rabi (winter) crop and is raised from October to April.

Similarly in other opium poppy growing areas all over the world, cultivars have been selected for production of maximum alkoids or oil from seed.

In addition to numerous garden hybrids the following varieties are recognized:

P. somniferum var. *glabrum* Boiss., cultivated in Turkey; flowers purplish but sometimes white; capsule subglobular; stigmata, 10-12; seeds white to dark voilet.

P. somniferum var. *album* DC, cultivated in India; flowers and seeds white; capsules more or less egg-shaped, 4-8 cm in diameter, no pores under the stigma.

P. somniferum var. *nigrum* DC, cultivated in Europe for the seeds, which are slate-colored and are known as 'maw seeds.' The leaves and calyx are glabrous, the flowers violet, and the capsules somewhat smaller and more globular than those of the var. *album.*

P. somniferum var. *steigerum* DC, a truly wild form found in southern Europe. The peduncles and leaves are covered with bristly hairs. The leaf lobes are sharply pointed and each terminates in a bristle.

Khanna and Gupta[10] observed that while there is a very rich diversity in the germplasm of Indian poppy no scientifically evolved variety is grown by the farmers. The material grown by the farmers is usually a mixture of various types.

Basing their information on the latest treatment of this subject by Asthana[4] collections were made of the cultivated poppy plants all over India.[10] A large collection of germ plasm has been evaluated and categorized. Khanna and Gupta's studies have revealed that there are no more than 20-25 basic cultivars which have been responsible for the existing diversity in the Indian germ plasm. A taxonomic key has been prepared by Khanna and Gupta for these cultivars on the basis of most salient features and some problems with regard to existing local names have been clarified. The performance of the major cultivars has been given with regard to opium yield and morphine content. The average opium yield (mg/plant) and morphine percentage for each of the cultivars observed by them is given in Table 2.3.

The occurrence of hetrosis has been studied in crosses between cv. Aphuri and 15 other cultivars. The maximum amount of heterosis over the superior parent is 46.34 percent for opium yield and 37.14 for morphine percentage.[11]

Chemical Races

Phillipson et al. reviewed species of section Miltantha of Papaveraceae for their contained alkaloids. Thebaine is a major alkaloid in *Papaver fugax* together with narcotine, pronuciferine, alpinigenine, O-ethylrhoeagnine, amurensinine, N-methyl O-crotonosine, armepavine, isocorydine, and salutaridine as minor alkaloids. Another sample of *P. fugax* contained glaucamine and glaudine as major alkaloids with rhoeadine, oreogenine, oreodine, and O-ethylglaucamine as minor alkaloids. Pronuciferine and armepavine were isolated as major alkaloids from a sample of *P. tauricola* with narcotine, roemerine, nuciferine, nantenine, and protopine as minor alkaloids. Another sample of *P. tauricola* yielded pronuciferine and mecambrine as major alkaloids with armepavine, lirinidine, thebaine, and cryptopine as minor alkaloids. A sample of *P. armeniacum* contained rhoeadine and rhoeagenine as major alkaloids with lirinidine, cryptopine, glaudine, and O-ethylrhoeagenine as minoralkaloids. Some 25 alkaloids representing nine different alkaloidal types were obtained from extracts of the three *Miltantha* species.

TABLE 2.3. Performance of some Indian varieties of opium poppy at Lucknow

S. No.	Parent	Source	Opium Yield (mg/plant)	Morphine Percentage
1.	Telia Kantiya	Neemuch	258	14.43
2.	Telia Kantiya	Mandsaur	259	13.64
3.	Ranjhatak	Neemuch	226	10.57
4.	Ranjhatak	Mandsaur	284	14.07
5.	Lal Kantiya	Mandsaur	233	12.27
6.	Telia	Mandsaur	217	14.06
7.	Liliya	Ratlam	295	11.25
8.	Doodhia	Barabanki	260	12.55
9.	Dhola Chotta Ghotia	IARI, N. Delhi	305	13.85
10.	Irranian	Gazipur	282	10.82
11.	Safed Dandi	Faizabad	282	12.66
12.	Bhakua	IARI, N. Delhi	325	12.95
13.	Kasuha	Mandsaur	249	15.47
14.	Dhaturia	Mandsaur	269	12.08
15.	Kantiya	N.B.R.I.	164	12.56
16.	Aphuri	Madhya Pradesh	114	11.46
C.D. at 5%			99.89	2.43
C.D. at 10%			134.53	3.27

This table was adapted from Khanna and Gupta.[10]

The results show that at least three different chemical races of *P. fugax* and *P. armeniacum* exist in which either 1-benzyltetrahydrori-soquin oline, proaporphine, aporphine, morphinane, or rhoedine types are the major alkaloids. There are at least two different chemical strains of *P. tauricola* which contain either 1-benzyltetrahydroisoquinoline, proaporphine, aporphine, or rhoeadine types as the major alkaloids.

Several species of Papaver *(P. persicum, P. acrochaetum, P. armeniacum, P. curviscapum, P. glaucum, P. macrostommu, P. bornmuelleri, P. rhoeas, P. dubium, P. hybridum).* Two species of Glaucium *(G. corniculatum, G. grandiflorum)* and *Roemeria hybrida* which are native to Iraq have been examined for their alkaloid contents by Phillipson et al.[24] These species belong to sections Miltantha, Papaver, Caratinae, Orthorhoedes, Argemonorrhoedes, Glaucium and Roemeria. The following 21 alkaloids were obtained from small samples which were mainly herbarium material: mecambrine, roehybrine floripavidine, N-methylasimilobine, roemerine, corydine, isocorydine, roemerin N-oxide, dehydroroemerine, liriodenine, alpinigenine, rhoeagenine, rhoeadine, glaucamine, glaudine, protopine, cryptopine, allocryptopine, cheilanthifoline, berberine and narcotine.

Section Oxytona

Goldblatt revised the section Oxytona (Macrantha) after intensive field studies coupled with cytological investigations and chemical analyses which resulted in the recognition of only three species viz. *Papaver bracteatum, P. orientaoe* and *P. pseudo-orientale.* These species could be distinguished not only on morphological characters but also some other characters.

a) *P. bracteatum* (diploid 2n = 14) major alkaloid is thebaine although some sample possess alpinigenine as a minor constituent.

It is reported that some strains of *P. bracteatum* Lindl produce high yields of thebain and therefore of commercial interest because thebaine can be converted easily into codeine.[30]

b). *P. orientale* (tetraploid 2n = 28) major alkaloid is oripavine with either thebaine or isothebaine as minor alkaloids.

c). *P. pseudo-orientale* (hexaploid 2n = 42) major alkaloid is isothebaine with either thebaine and/or oripavine as minor alkaloids.

Phillipson[26] reported that at least three chemical races of *P. bracteatum* are known to exist, one containing thebaine (Morphine-type) a second thebaine with alpinigenine (rhoeadane-type) and a third with thebaine and orientalidine (tetrahydroberberine-type).[27] Similary, *P. orientale* exists in six different chemicals races with oripavine, oripavine and thebaine (morphinan-type), oripavine and isothebaine (aporphine-type), oripavine and alpinigenine (rhoeadane-type), oripavine and alpinigenine (rhoeadane-type), oripavine and alpininegine and thebaine and macambridine (tetrahydroberberine-type) as major alkaloids.[28]

P. pseudo-orientale also exists at least in four different chemical races, one containing isothebaine (aprorphine-type) as major alkaloid (2n = 42) with varying quantities of mecambrindine and orientalidine (tetrahydroberberine-type), a second containing salutaridine (promorphinane-type) and thebaine as major alkaloids (2n = 14), a third containing salutaridine as major alkaloids with isothebaine, mecambridine as minor alkaloids (2 = 28) and a fourth containing macrantaline (secoberberine-type) and salutaridine as major alkaloids (2n = 14).

It is assumed at this stage that the alkaloids are controlled genetically and that chemical races exist. In order to confirm it is necessary that plants of known alkaloids composition be grown under different conditions (Phillipson).[26]

An Iranian sample of *P. psuedo-orientale*[29] exhibited isothebaine as major alkaloid along with the presence of salutaridine, bracteoline, orientalidine, arapavine, alborine, PO4, Or_1 and Or_2.

The Turkish species of Oxytona are represented by *P. orientale* and *P. pseudo-orientale* only since *P. bracteatum* has not been found in Turkey. A turkish sample of *P. pseudo-orientale* yielded salutaridine and macrantaline as major alkaloids with macrantoridine as a minor alkaloid.[23,27]

The alkaloids obtained from the capsules of five different collections of wild *P. orientale* and 16 collections of wild *P. pseudo-orientale* have been investigated and chromosome numbers of some of the samples determined.[28] Oripavine was the major alkaloid obtained from four of *P. orientale* collections, two of which had a diploid chromosome number of 28. These four samples contained isothebaine and alpinigenine as minor alkaloids whereas two contained mecambridine and orientalidine and the two contained thebaine and salutaridine. The remaining sample of *P. Orientale* contained mecambaridine as the major alkaloid and also some orientalidine.

Thirteen of the sixteen samples of *P. pseudo-orientale* contained isothebaine, mecambridine and orientalidine as their major alkaloid with thebaine and salutaridine as minor alkaloids. Alpinigenine was detected in three samples. The diploid chromosome number of four of the thirteen samples was determined as 42.

Two samples of *P. pseudo-oreintale* contained salutaridine and thebain as major alkaloids and had a diploid chromosome number of 14, whereas the remaining samples yielded salutaridine as the major alkaloid with isothebaine, mecambridine and orientalidine as minor alkaloids and a diploid chromosome number of 28.

REFERENCES

1. Anonymous. Wealth of India, Raw Materials, Vol. VII. Council of Scientific Industrial Research, New Delhi, 1966, pp. 231-246.

2. Fedde, F. Papaveraceae. In Die Naturfichen Pflanzenfamilien (ed. Engler, A. and Prantl, K.) 2. Augl. 17B, 1936, p. 59.

3. Trease, G. E. and Evens, W. C. Pharmacognosy-12th Edition. Bailliere Tindall, London, 1983.

4. Asthana, S. N. Cultivation of Opium Poppy in India. Bull Narcotics, Sept.-Dec., 1954, pp. 1-10.

5. Veselovskaya, M. A. The Poppy. Amerind Publishing Co., New Delhi, 1976. (Translated from Russian.)

6. Ramnathan, V. S., and Ramchandran, C. Opium Poppy Cultivation, Collection of Opium, Improvement and Utilization for Medicinal Purposes. In Cultivation and Utilization of Medicinal and Aromatic Plants (ed. Atal., C. K. and Kapoor, B. M.). R. R. L. Jammu Tawi, 1977, pp. 38-74.

7. Private Communication-Curator Indian Botanic Gardens Calcutta.

8. Hocking, G. M. A Dictionary of Terms in Pharmacognosy. C. C. Thomas, Publisher, Springfield, Illinois, 1955.

9. Kirtikar, K. R., Basu, B. D. and I. C. S. (retired). Indian Medicinal Plants, Vol. 1-4, Platas 1-4. L. M. Basu, Publ. Allahabad, 1918; 2nd ed., 1935. Reprinted Bishensingh, Mahinderpalsingh Dehra Dun, 1975.

10. Khanna, K. R. and Gupta, R. K. An Assessment of Germ Plasm and Prospects for Exploitation of Heterosis in Opium Poppy (*Papaver somniferum*). Con-

temp. Trends in Plant Sciences 1981 (ed. S. C. Verma), Kalyani Publishers, New Delhi, 1981.

11. Anon. NBRI Newsletter, 1988 XV (1) National Botanical Research Institute, Lucknow.

12. Phillipson, J. D., Thomas, O. O., Gray A. I. and Sariyar, G. Alkaloids from *Papaver armeniacum P. fugax* and *P. tauricola*. Planta med., 1981, 41, p. 105.

13. Hutchinson, J. The Families of Flowering Plants. Vol. I, Reprint ed. Macmillan & Co., London, 1960.

14. Stermitz, Frank R. Alkaloid chemistry and the systematics of Papaver and Argemone in "Recent advances of Phytochemistry." Vol. I, Ed. T. J. Mabry, R. E. Alston & V. C. Runeckles. Appleton-Century-Crafts, New York, 1968.

15. V. Preininger. Chemotaxonomy of Papaveaceae and Fumariaceae. In Alkaloids ed. Arnold Bross Vol. 29. 1986. Academic Press Orlando.

16. J. J. Bernhardi, Linnaea 8. 401, 1833.

17. L. Elkan "Tenatamen monographiae generis papaver." Regimentii Borossuorum Konigsberg 1839.

18. K. Prantl and L. Kinding in "Die naturlichen Pflanzenfamilien" (A. Engler and K. Prantl, eds.) Part III/2, p. 130. Engelmann, Leipzig, 1889.

19. F. Fedde. In "Das Pflanzenreich" (A. Engler, ed.) Part IV, No. 40, p. 288. Engelmann, Leipzig, 1909.

20. K. F. Gunther, Flora (Jena) 164 (4), 393, 1975.

21. J. Novak and V. Preininger, Preslia 52, 97, 1980.

22. V. Preininger, J. Novak, and F. Santavy, Planta Med. 41, 119, 1981.

23. Sariyar, G. J. Fac. Pharm. Istanbul 12. 171, (1976).

24. Phillipson, J. D., Gray, Alexander I., Askari, Ali A. R., Khalil, Afafa. Alkaloids from Iraq species of Papaveraceae. Jour Nat. Prod. Vol. 44, No. 3, (1981) 296-307.

25. Goldblatt, P. Annals of Missouri Botanic Garden 61:265, (1974).

26. Phillipson, J.D. Infraspecific, variation and alkaloids of Papaver species. Planta medica vol. 48, pp. 187-192, 1983.

27. Sariyar, G. and Phillipson, J.D. Phytochemistry 16, 2009. 1977.

28. Phillipson, J. D., Scutt, A., Baytop, A., Ozhatay, N., and Sariyar, G. Alkaloids from Turkish samples of Papaver orientale and P. pseudo-orientale. Planta Medica *43*, pp. 261-271, (1981).

29. Shafiee, A., Lalezari, I., Nasseri-Nourri, P. and Asgharian, R. J. Pharm. Sc. 64, 1570, (1975).

30. Nyman, U. and Bruhn, J. G. Planta Medica 35, 97, (1979).

Chapter 3

Cytology and Genetic Studies

The genus *Papaver* is cosmopolitan in distribution.[1] It includes 99 or more species which invariably contain milky latex in specialized cells or ducts called laticifers. Almost all the species contain one or the other alkaloids such as morphinane, promorphinane, 1-benzyltetrahydroisoquinoline, proaporphine, tetrahydroprotoberberme, protopine, 1-benzylisoquinoline, rhoeadane, isopavine type of alkaloids.[2]

Two species of *Papaver,* viz *P. setigerum* and *P. somniferum,* need special reference as they contain one of the most important alkaloids, morphine, in their latex and edible oil in their seed.

P. setigerum occurs in a wild state in the South Mediterranean region and in the Canary Islands. *P. somniferum* is not reported to occur in a wild state but is cultivated for opium production or oil seeds in Turkey, Afghanistan, Iran, Pakistan, India, Burma, Laos, Thailand, China, Germany, France, Italy, Canada, Mexico, Bulgaria, the former Yugoslavia, the former USSR, Argentina, Ecuador, Peru, and Panama. It is a long day plant and cross-pollination occurs in the cultivated population. Automatic self-pollination and self-fertilization also occurs in *P. somniferum.*[3]

The flowering of the plant begins with formation of buds on the main stem, which is followed by flowers on the lateral axes which gradually begin to bloom in the morning. On dry sunny days the flowers open even before the sun rises, but they open much later when the weather is wet. Similarly, the falling of petals depends on humidity and temperature. Some experiments in the former USSR prove the fertility of poppy as a result of self-pollination and the possibility of its being pollinated in the bud stage.[4]

In Russia and in the adjoining east European countries genetic and breeding studies have been carried on by all conventional methods with considerable success for the increased yield of morphine or other alkaloids and increased yield of seeds for oil per hectare.

Little or no work on plant breeding studies has been done in India until recently whereas *P. somniferum* has been grown on quite an extensive scale since the fifteenth century. The cultivation, possession, and trade of opium poppy has been the monopoly of the federal government and any violation is an offense. This restriction, though necessary for other reasons, did not create a congenial atmosphere for scientists to initiate crop improvement research work. However, scientists under the Council of Scientific Industrial Research, Indian Council of Agricultural Research, and some universities have recently initiated a coordinated research program for a crop improvement and the results are encouraging.

Morphine contents of opium poppy vary from 10 to 12 percent in India as compared to over 20 percent in eastern Europe and Russia. Musalevski and Teodosievski[5] reported 9.53 to 22.42 percent morphine in come cultivars and in Japan the opium harvested during 1981 to 1985 showed 8.03 to 17.2 percent morphine.[6]

There is good potential to improve the morphine contents of opium poppy in India, which has a cultivated crop with high genetic variability and higher percentage of morphine contents achieved in other countries.

For any crop improvement program or to study the phylogeny, it is necessary to know the basic chromosome numbers of the plants.

In Fedde's monograph,[7] the fourth section of the genus, *Mecones*, includes five species, viz. *P. somniferum L.*, *P. setigerum DC*, *P. glaucum Boiss*, *P. gracile Auch*, and *P. decaisnei Hochst*. Leger[8] observed that *Papaver* is the most highly developed genus in the Papaveroideae.

This genus has about 90 species or more, which are grown mostly in the Old World, though a few are native to the western area of North America. Unfortunately, rather a small number of karyological studies are to be found, although most of the species are common as garden plants.[9]

Karyological studies are made of this genus shown in the following list reproduced from Sugiura's phylogenetic studies in papaveraceae:

Tahara (1915) on *P. rhoeas, somniferum,* and *orientale*; Ljungdahl (1922) on *P. atlanticum, lateritium, persicum, tauricolum, hybridum, nudicaule, alpinum, rhoeas, pilosum, dubium, orientale, striatocarpum, radicatum,* and *somniferum*; Yasui (1921-1937) on *P. orientale, somniferum, bracteatum,* and *lateritium*; Vilcins and Abele (1927) on *P. rhoeas*; Philip (1933) on *P. commutatum*; Lawrence (1930) on *P. rhoeas*; Kuzmina (1935) on *P. glaucum, setigerum.*

Sugiura[9] observed meiotic chromosome numbers n = 7 for *Papaver glaucum,* n = 11 for somniferum var. danebrog and n = 11 for *P. setigerum.* Earlier Tahara and Yasui observed n = 11 for *P. somniferum* and Ljungdahl observed n = 22 for *P. setigerum.* Kuzmina[11] had recorded "It is this wild growing species (*P. setigerum*) which is almost unanimously recognized by botanists as the progenitor of cultivated poppy (*P. somniferum*)."

But Sugiura[9] was of the opinion that *P. glaucum* and *P. gracile* are the progenitors of *P. somniferum.* He based his explanation that *P. glaucum* has seven meiotic chromosomes consisting of 4 + 3; *P. somniferum* has 11 meiotic chromosomes consisting of 4 + 4 + 3 = 11, where 4 + 3 are the basic chromosome numbers of Papaveraceae.

Yasui[10] (1936-1937) also gave the chromosome constitution of *P. somniferum* (n = 11) as $4_{11} + 3_1$ genetically in her studies.

Up to this time, two basic numbers have been found in the genus *Papaver* but it has not been explained why so. According to Sugiura, Darlington treated these numbers as uneven multiples. Sugiura,[9] after his exhaustive study of Papaveraceae, concluded that Papaveraceae, Argemone, Meconopsis, and Roemeria are derived from both Corydaleae and Chelidonieae, the former having the basic number 4 and the latter 3. Thus the meiotic chromosome number 7 which is generally found in those genera of Papaveraceae would be the sum of 4 and 3 (amphidiploidy). In the same way the number 11 found in section Mecones consists of 4 + 4 + 3. The same constitution of chromosomes was also explained from a genetical point of view by Yasui.[10] The meiotic chromosome number 11, being found only in section Mecones, is of special interest.

Some of the data obtained for the plants of genus *Papaver* with regard to number and size of chromosomes are tabulated in Table 3.1.[9]

Based on chromosome numbers, it is evident there are two basic

TABLE 3.1. Number and size of chromosomes in some species of *Papaver*

Plants Investigated	n	IM(μ)	Chromosome Length IA(μ)	IIM(μ)
Papaver hybridum	7			0.6
P. hookerii	7	1.15		
P. somniferum				
V. danebrog	11	1.75		
P. setigerum	11			1.25
P. armeniacum	7	1.75		
P. caucasicum	6(5+L)		1.0;1.25	
P. desertorum	7		1.125	
P . floribundum	7	1.3		
P. persicum	7			0.65
P. atlanticum	7(6+S)		0.9;0.37	
"	6(5+L)	0.9;1.5		
P. heldreichii	7(6+S)	1.4;0.9		
P. lateritium	7			1.0
P. schinzianum	7(6+S)		1.125;0.45	
P. strictum	7(6+S)	1.1;0.75		
P. burseri	7	1.1		
P. radicatum	7	1.125		

*Sugiura, T. 1940[9]

numbers in the Papaveraceae, one 4, which is found in Hypecoideae and Fumariodeae and the other 3, which is found in Pteridophyllaceae and Papaveroideae except *Papaver, Argemone, Meconopsis,* and *Roemeria.* The basic number 7 is found in Papavereae, excluding *Glaucium* and some *Roemeria.* The secondary basic number 11 is often found in a section of *Papaver* (Mecones) and some Roemeria. *Papaver* can be said to be the most highly developed genus among Papaveraceae, both for anatomical and karyological reasons and has become the most advanced genus in Europe and Asia Minor, having about 90 or more species.[9]

Fedorov[12] has compiled the list of plants which have been investigated for chromosome numbers up to 1967. Plants of *Papaver* investigated so far by different workers are tabulated in Table 3.2. For every species, 2n chromosome numbers are given along with the authority and the year of investigation.

Koshy and Mathews[13] reported the following meiotic chromosome numbers of the following Papaver species from South India:

Papaver rudicaule – n = 14
P. rhoes – n = 7
P. somniferum – n = 11

Kaul et al.[26] observed that the genome of *P. somniferum* consisted of two pairs of very long (10.27-11.40 μm.), two pairs of long (8.65-9.37 μm.) ,four pairs of medium (7.50-7.87 μm.), and three pairs of short (6.37- 6.75 μm.) chromosomes. All were subterminal except the chromosome number 8 which was submedian.

Floria and Ghiorghita[27] examined mitotic behavior of seedling grown from chemically treated seeds. Chromosomal aberration during anaphase and telophase of root meristem occurring in higher +frequency than in the control was observed by them. The presence of fragments, bridges, laggards, and micronuclei was also reported by them. Gohil and Kaul[28] reported structural hybridity in 2 species of Papaver. Multivalents were observed in *P. somniferum* and *P. rhoeas*. Some plants of *P. rhoeas* exhibited moderately strong desynapsis.

Patra et al.[29] reported that cytomixis is a practically manoeuvered process for the generation of syncytes in *P. somniferum*. They concluded that higher ploidy level coupled with less sterility may be instrumental for variation and evolution in opium poppy.

The studies by Farmilo et al.[14] also showed that *P. somniferum* is diploid (n = 22) while *P. setigerum* is tetraploid (n = 44). No other species of Papaver is known to have the number n = 11 or multiple of 11. This indicates that *P. setigerum* cannot be the ancestoral species of *P. somniferum* since a diploid would not be expected to be derived from a tetraploid.

Hammer and Fritsch,[15] however, observed that diploid forms of subspecies *setigerum* may be the progenitor of diploid *P. somniferum* subspecies somniferum. In their opinion setigerum and somniferum are the two subspecies of *P. somniferum*. Veselovskaya[4] proved that the two species are closely related, as she obtained reproducible hybrid seeds from crosses between *P. setigerum* and *P. somniferum*. Since they differ in their morphological characters they are considered independent species with common ancestral origin. Hrishi[16] reported fairly good genomic affinity between two species. He observed good pairing of the chromosomes belonging

TABLE 3.2. Chromosome numbers of flowering plants

Species	Chromosome na.	Author studied by
Papaver L.		
alboroseium Hult	28	Knaben 1959a, b
alpinum L.	14	Ljungdahl 1922
		Sugiura 1936a, 1937a
		Faberge 1942, 1943, 1944
		Knaben 1959a, b
		Ernst W. R. (D. n. 1965)
amurense N. Busch	40-42	COKOJIOBCKAR1966
apulum	12	Sugiura 1936a, b
		McNaughton 1960
argemone L.	12	Beal (Maude 1939)
	40	Koopmans 1954
		(L. 1961)
	42	Sugiura 1936a, 1937a
		Rohweder 1937
		Kawatani, Ohno (Asahina
		et al. 1957)
		McNaughton 1960
armeniacum	14	Sugiura 1938
atlanticum (Ball.) Coss.	14	Ljungdahl 1922
		Sugiura 1937c
		Ernst W. R. (D. n. 1965)
bracteatum Lindl.	14	Yasui 1936b, 1937a, 1941
	42	Kawantani, Obno
		(Asahina et al. 1957)
		Ernst W. R. (D. n. 1965)
burseri	14	Sugiura 1937c
californicum A. Gray	28	Ernst W. R. 1958
carmelii Feinbr.	14	Feinbrun 1963
caucasicum	12	Sugiura 1937b
chibinense	14	Matbeeba, Thxohoba
		Heohyoji.
commutatum Fisch.		
et Mey.	14	Philp 1933b
		Sugiura 1936a, 1937a
cornwallisense D. Love	84	Love A. 1962c
dahlianum Nordh.	70	Horn 1938
		Knaben 1958, 1959a, b
		Love A. Love D. 1961b
		Love A. 1962c

Species	Chromosome na.	Author studied by
desertorum	14	Sugiura 1937c
dubium L.	14	Sugiura 1936a
	14, 28	Rohweder 1937
	28	Ljungdahl 1922
		Tischler 1934
	42	Sugiura 1937a
		Love A., Love D. 1944b
		McNaughton 1960
		McNaughton, Harper 1960
		Heimburger C.
		(L. 1961)
		Gadella, Kliphuis 1966
fauriei Fedde	16	Kawano 1963a
floribundum	14	Sugiura 1938
fugax	14	Sugiura 1940d
glaucum	14	Sugiura 1931, 1936b, 1937b
		KY3LMHH 1935
heldreichii	14	Sugiura 1937c
hookeri	14	Sugiura 1937b
hultenii Knaben	42	Knaben 1959a, b
humile Fedde		
subsp. sharonense Feinbr.	14	Feinbrun 1963
hybridum L.	14	Ljungdahl 1922
		Sugiura 1937c
		Kawantani, Ohno
		(Asahina et al. 1957)
		McNaughton 1960
kerneri Hayek	14	Faberge 1943
laestadianum Nordh.	56	Horn 1938
lapponicum (Tolm.) Nordh.	24	COKOJIOBCKAR,
		CTPEJIKOBA 1960
	56	Horn 1938
		Nygren (Love A.,
		Love D. 1948)
		Knaben 1958, 1959a, b
lateritium C. Koch	14	Ljungdahl 1922
		Sugiura 1940c (L. 1961)
		Yasui 1941
		Kawatani, Ohno (Asahina
		et al. 1957)

TABLE 3.2 (continued)

Species	Chromosome na.	Author studied by
lecoqii Lamotte	28	McNaughton 1960
	42	Heimburger C. (L. 1961)
litwinowii Fedde ex Huet	14	Matbeeba, Thxohoba Heohyoji.
macounii Greene	28	Horn 1938
		Knaben 1959a, b
nordhagenianum A. Love	70	Love A., Love D. 1948
		Love A. 1954b, 1955, 1962c
nudicaule	14	Ljungdahl 1922, 1924
		Yasui 1927
		Sugiura 1940c (L. 1961)
		Faberge 1942, 1944
		Fagerlind 1944
		Knaben 1959b
		Love A., Love D. 1961b
	28	Horm 1938
		Ernst W. R. (D. n. 1965)
var. striatocarpum Fedde	70	Ljungdahl 1922, 1924
oreophilum	14	Sugiura 1936a, 1937a
orientale L.	28	Snoad 1952 (D. 1955)
	42	Tahara 1915f
		Yasui 1921
		Ljungdahl 1922
		Castiglia 1955
		Kawatani, Ohno (Asahina et al. 1957)
		Kawatani, Asahina 1959
pavoninum	12	Sugiura 1931, 1936b
persicum	14	Ljungdahl 1922
		Sugiura 1937c
pilosum	28	Ljungdahl 1922
pinnatifidum Moris	28	Sugiura 1936a, 1937a
		Feinbrun 1963
pinnatifolium	24	Sugiura 1936a, b
pseudo-aussknechtii Fedde	14	Feinbrun 1963
pygmaeum Rydb.	42	Faberge 1944
		Knaben 1959b
pyrenaicum (L.) Kern.	14	Faberge 1943
		Knaben 1959a
radicatum Rottb.	14	Sugiura 1940c
	28, 56	Love A. 1962b
	42, 56, 70	Faberge 1944

Species	Chromosome na.	Author studied by
	56	Holmen 1952
		Love A. 1954b,
		1955a, 1962c
		KYKOBA 1965a
	56, 70	Horn 1938
		Knaben 1958
		Mosquin, Hayley 1966
	56, 70, 84	Knaben, Jorgensen
		(Jorgensen et al. 1958)
	70	Ljungdahl 1924
		Flovik 1940
		Faberge 1942
		Love A., Love 1948
		Knaben 1959a, b
relicium (Lundstr.) Nordh.	70	Horn 1938
rhoeas L.	14	Tahara 1915f
		Ljungdahl 1922
		Vilcins 1927
		Lawrence 1930
		Tischler 1934
		Yamazaki 1936
		Rohweder 1937
		Sugiura 1940c (L. 1961)
		Felfoldy 1947a
		Castiglia 1955
rhoeas L.	14	Hasitschka 1956
		Kawatani, Ohno (Asahina
		et al. 1957)
		McNaughton 1960
		Mitra K. 1964
		Ernst W. R. (D. n. 1965)
	14, 15, 21	Koopmans 1956
rubro-aurantiacum		
(Fisch.) Lundstr.	28	Horn 1938
		Knaben 1959a, b
rupifragum Boiss. et Reut.	12	Sugiura 1936a, b
	14	Snoad 1952 (D. 1955)
		Quezel 1957
schinzianum	14	Sugiura 1940c
		(D. 1945)
sendtneri Kern.	14	Faberge 1943

TABLE 3.2 (continued)

Species	Chromosome na.	Author studied by
setigerum DC.	22	Sugiura 1938
	44	Ljungdahl 1922
		KYELMHHA 1935
		Hrishi 1960
somniferum L.	20	Ghimpu 1933
	22	Tahara 1915f
		Ljungdahl 1922
		Yasui 1927
		KYELMHHA 1935
		Sugiura 1940c (L. 1961)
		BOJIOTOB 1941
		Castiglia 1955
		Kawatani, Ohno (Asahina et al. 1957)
		Kawatani, Asahina 1959
		Hrishi 1960
		Ernst W. R. (D. n. 1965)
stefanssonianum A. Love	70	Love A. 1955
steindorssonianum A. Love	70	Love A. 1955
striatocarpum	70	Ljungdahl 1924
strictum	14	Sugiura 1937c
suaveolens Lap.	14	Kupfer, Favarger 1967
subpiriforme Fedde	14	Feinbrun 1963
syriacum Boiss. et Blanche	14	Feinbrun 1963
tauricolum	14	Ljungdahl 1922
tenuissimum (Heldr.) Fedde	14	Feinbrun 1963
trinifolium L.	28	Sugiura 1937c
umbrosum	14	Sugiura 1931, 1936b
sp.	42	Knaben 1959b
sp.	84	Knaben 1959b
sp.	42, 56	Knaben 1959a, b

Adapted from *Chromosome Numbers of Flowering Plants*. A. A. Fedorov. 1969.

to a genome in both the species, *P. somniferum* (2n) and *P. setigerum* (4n). More or less this confirms the conclusion drawn by Hammer and Fritsch[15] that *P. somniferum* seems to have originated most probably from the diploid *P. setigerum* through the allopatric evolutionary pathway. But Sugiura[9] hesitates to recognize *P. setigerum* as a direct ancestor of *P. somniferum*, thinking it more correct to regard *P. somniferum* and *P. setigerum* as independent species though genetically near species having evidently arisen from one ancestor.

The cultivated species *P. somniferum* has two subspecies, viz. somniferum and hortense. The former bears indehiscent capsules of varying sizes containing latex when immature and white seeds when matured. The subspecies hortense bears dehiscent capsules with no opium but it is useful for seed production.

The opium containing subspecies *P. somniferum* further includes two varieties, viz. var. *glabrus* and var. *album*. The var. *glabrus* is grown mostly in European countries for seed and oil, the var. *album* is cultivated in India and other countries for opium production.

Genetic variation has been observed in the cultivated population of *P. somniferum* as expressed in color of petals, shape and size of the capsule, floral deviation, percentage of morphine, and other alkaloidal contents. Sharma[17] reported discrete phenological variation for six qualitative traits and four meteric traits, scored qualitatively occurring naturally even in a limited genetic stocks obtained from Uttar Pradesh, Madhya Pradesh, and Rajasthan states of India. Gupta et al.[18] reported white, pink, deep red, and purple-red colors of petals with smooth or fringed margins and also an exciting range of additional colors like purple, blue, and a blend of two or more colors. In Uttar Pradesh most of the cultivated poppy consists of white petals and white seed but in Madhya Pradesh and Rajasthan the petals have other shades.

Roelofs[19] reported occurrence of 'eversporting' varieties where secondary pistils (*polycephaly*) are produced in the flower as a case of abnormal variant. Ilieva et al.[20] observed a mutant tetraploid form of Bulgarian opium poppy. Many spontaneous mutants with altered floral structures were also observed by many research workers.[21]

Musalevski and Teodosievski[5] found that morphine content of the crude opium ranged from 9.53 to 22.42 percent with four populations exceeding 19 percent morphine. Voskerusa[22] found genetic and environmental variations for morphine content in 64 varieties. He indicated that south European varieties have the least amount of morphine but Coiciu et al.[23] reported that Rumanian varieties had satisfactory contents of morphine and also had a good amount of seed and yield of oil.

The amount of opium alkaloids, viz. morphine, codeine, and narcotine, in the capsule varied greatly when harvested at biological maturity. Thebaine and papaverine were the highest at ripeness for processing; but the total alkaloid content was highest (0.858 percent) at biological maturity.[24]

Nymann and Hansson[25] reported that generally higher morphine content was observed in varieties where a type of the alkaloid phthalideisoquinoline was absent. They were of the opinion that improvement in percentage of the morphine may possibly be achieved by eliminating this type of alkaloid.

Breeding of poppy was intensified, but mostly in the former USSR and adjoining east European countries. The major approaches were simple selection, hybridization, and exploitation of hybrid vigor for production of poppy seed. Improvement of this crop has been initiated recently in India.

HETEROSIS IN HYBRID

Khanna and Gupta,[30] after collecting a large number of germ plasm in India, concluded that there are about 20 basic types which have given rise to wide diversity by intermixing and hybridization. The occurrence of heterosis is frequent and is of considerable magnitude but the genotypes whose hybrids would give better performance have to be identified and fixed, although enough data has been collected for developing the concept of plant type in opium poppy.

There are reports on the occurrence of heterosis for latex yield, seed yield, morphine content, plant height, and number of nodes, etc., in inter-inbred or inter-varietal hybrids. In a few studies morphine content was intermediate between the parental values. Sarkany et al.[32] developed some inter-inbred hybrids which gave better performance in length of stem, early flowering, number of capsules, and seed yield. Morphine content was also higher in many cases. Michna and Szwadiak[33] in the crossing experiments also observed higher seed yield and higher morphine content. It was generally observed that hybrids manifested heterosis for capsule weight and morphine percentage.

Singh and Khanna[34] observed significant marked heterosis for capsule number and opium yield over the superior parents. Kaicker et al.[35] also found some crosses better for opium yield, capsule size, capsule number, and positive heterosis for morphine percentage. Saini and Kaicker[36] obtained maximum heterosis for capsule number and minimum for flowering time when they crossed exotics with indiginous strains of *P. somniferum.*

Singh and Khanna[34] observed marked heterosis for all the characters in a set of 6-parent half-diallel for opium yield, seed yield, oil content, and eight other component traits. Capsule/plant (46.90 percent) showed the maximum heterosis, followed by branches/plant (44.96 percent), opium yield (40.77 percent), and seed yield (28.85 percent).

Khanna and Gupta[30] reported the occurrence of heterosis in crosses between cv. Aphuri and 15 other cultivars of opium poppy. The maximum amount of heterosis over the superior parent was 46.34 percent for opium yield and 37.14 percent for morphine percentage. In Bulgaria, Popov et al.[37] developed new cultivars of *P. somniferum* which are more cold resistant and produce higher yield of dry capsules with higher morphine contents by using generically remote poppy forms from *P. turcicum* and *P. eurasiaticum.*

Genetic studies and economic potential of interspecific crosses between the two species, viz. *P. somniferum* and *P. setigerum,* containing morphine, the principle opium alkaloid, have been reported.

Vigorous triploids between the two species evolved but the extent of triploid heterosis for yield and contributing characters had not been fully determined.

P. setigerum has several desirable features such as high capsule number and high papaverine contents which are worth introducing in *P. somniferum.* With these objectives in view hybridization between the two was undertaken and studied up to the F4 generation.[38]

F_1 generations are vigorous, tall, and generally bear nine to 23 capsules per plant (average 14.75). The capsules are larger in size (1.60 cm $\times$ 2.12 cm) = 3.39 cms^2 on an average than in *P. setigerum* but has the same shape. The plants are invariably triploid, the average number of bivalents and univalents being 11 each. Some plants yielded 560 mg of dry opium which comes to potential yield of 100 kg/ha. Twenty-five percent plants yielded more than 400 mg opium/plant which would be equivalent to about 80 kg opium/ha.

Even the average yield of 290 mg/plant, equivalent to 60 kg of dry opium/hectare would surpass some of the best cultivars of *P. somniferum*. The seed set was poor in F_1 (1-4 g), average 1.9 g seed produced per plant. The F_2 showed a marked segregation for all the morphological characters. The plants are mostly triploid aneuploids and devoid of heterotic effect of opium yield as seen in F_1. Average opium yield 148.215 mg/plant or 30 kg/ha with no improvement in seed production was observed. In F_3 the opium yield falls further but the capsule size and seed production show improvement. An average seed yield is two to six g/plant.

In F_4 there is general reduction in number of capsules in comparison to F_3 but on an average eight to nine capsules per plant have been observed. The seed yield was 4-12 g/plant and opium estimated yield 92 kg/ha.

Khanna and Shukla[39] reported the inheritance of five major alkaloids, viz. morphine, codeine, thebaine, narcotine, and papaverine, in the interspecific crosses of *P. somniferum* and *P. setigerum*. Heterotic increase in codeine and thebaine was found in some F_2 plants. But in F_2 plants contents of these alkaloids except that of codeine exceeded the contents found in the parental and F_1 generation. The absence of narcotine was generally dominant over its presence.

Triploid heterosis at F_1 has been observed for the thebaine and transgressive segregation was observed in F_2 for morphine, thebaine, and papaverine. The alkaloid analysis in the subsequent generations has shown that high concentrations observed in the earlier hybrids could not be maintained. High papaverine content has been correlated with white fresh latex in contrast to the brown or red latex normally observed in *P. somniferum* cultivars. On this basis high papaverine (up to 6 percent in some individuals) lines have been isolated.

MALE STERILITY

Singh and Khanna[40] observed male sterile plants both spontaneous as well as in material irradiated with 10 and 20 kr from a cobalt gamma source. When these were crossed with normal plants it produced male sterile types in subsequent generations (F_1 and F_2)

indicating a complex pattern of inheritance of male sterility. Hirishi[16] also observed male sterility in F_2 generation of interspecific hybrids between *P. somniferum* and *P. setigerum*. He also reported that male sterility (pollen sterility) was correlated with flowers' color.

MUTATION BREEDING

Khanna and Singh[41] used 10 and 20 kr of gamma rays to irradiate opium seeds and found some economically viable mutants of three types:

(a) male sterility of different kinds
(b) opiumless mutants
(c) high morphine yielding plant (20.6 percent of morphine)

Under the influence of ionizing radiation Ilieva et al.[42] also recorded biological and biochemical changes.

The mutants thus evolved might prove very useful for creation of novel genotypes for improved chemical compounds or for morphological characters.

POLYPLOIDY BREEDING

Induction of polyploidy has been attempted in opium poppy by various workers. Furusato[43] obtained tetraploid plants of *P. somniferum* by colchicine. Andreev[44] recorded a triploid heterosis for both morphine contents and yield of opium in a trial consisting of diploid, triploid, and tetraploid forms of the same variety of *P. somniferum*.

Polyploidy, especially triploidy, seems to hold promise of commercial viability in opium poppy.

Trease and Evans[45] reported that morphine yield per unit area increased up to 100 percent in polyploids especially in 3n plants.

By the irradiation of poppy seeds with ^{60}Co a number of mutations have been produced, including ones affording plants with increased morphine content; these increases were maintained in X_2 generation with an average morphine content of 0.52 percent compared with 0.32 percent for the controls.[45]

The successive treatments of *P. somniferum* seeds with gamma rays and alkylating agents in the first generation and gamma rays in the second generation induced a high variability of the plants. The self-pollinated poppy plants had a clearly high morphine content in the capsules. There were isolated individuals of *P. somniferum* with relatively high morphine content in the capsules (0.53-0.54 percent dry wt.).[46]

Production of Morphinan Alkaloids by Means of Cell Cultures

Verpoorte et al reported that significant amounts of morphinan alkaloids were found only in 0.15% codeine in a cell suspension culture of *P. somniferum*[65] and 4.7% in a callus and 5.6% in a cell suspension culture of *P. somniferum*.[66] In the latter case all six major opium alkaloids were reported to be present in the cultures, morphine, narceine, and noscapine being the major alkaloids (respectively 1.6, 1.9, and 1.8%).

In callus cultures of *P. somniferum* it was found that growth hormones affect the total amount of morphinan alkaloids formed. The ratio between the two alkaloids formed, thebaine and codeine, was dependent on the auxins added.[67] Hodges and Rapoport[65] detected morphinan alkaloids by means of radioimmunoassay in callus cultures of *P. somniferum*. However, after repeated subculturing the callus cultures ceased production of Morphinan alkaloids. Heinstein[69] reported a 10 to 20-fold increase in production of morphinan alkaloids in *P. somniferum* cell suspension cultures after elicitation with sterilized *Verticillium dahlia* or *Fusarium moniliforme* conidia resulting in yields of approximately 4 to 5 mg/liter of both codeine and morphine. Kamimura and co-workers[70-72] reported extensively on the optimization of growth of callus and cell suspension cultures of *P. bracteatum*. The cultures contained low levels of thebaine; however, on prolonged subculturing the levels decreased considerably to only trace amounts of alkaloids. By subculturing a cell suspension on an auxin-free medium promoting aggregation, thebaine levels could be increased.

From the results with various type of cultures it can be concluded that the content of morphinan alkaloids increased with the degree of differention (e.g.,[67,73,74,75,76]). On media-inducing root[67,74] or embryo formation in *P. somniferum* small amounts of morphinan alkaloids could be detected. In *P. bracteatum* morphinan alkaloids are also produced in root and embryo cultures.[74,77] Nessler and Mahlberg[78] reported that root and shoot cultures induced from callus cultures had cells similar to laticifers in intact plants. Kutchan et al.[50] found a correlation between the occurrence of such laticifer-like cells and production of morphinan alkaloids (thebaine) during cytodifferentiation of cultured cells of *P. bracteatum*. The cytodifferenciation characterized by the formation of shoots was induced by removal of hormones from the culture medium. A correlation was also found in seedlings between the appearance of morphinan alkaloids and laticifer cells.[81] Differentiation of poppy cell culture into shoot-forming meristemoids results in the formation of morphinan alkaloids as well.[73] Again the presence of trace elements and laticifers in the calli were thought to be a prerequisite for morphinan alkaloid production. Czygan and Abou Mandour[52] reported the occurrence of thebaine in *P. bracteatum* callus cultures. No laticifer cells could be detected, and no other form of differentiation could be observed. However, alkaloid levels were considerably lower than reported for differentiated cultures containing laticifer. Griffing et al.[75] used a radioimmunoassay for the analysis of morphinans in *P. somniferum* hypocotyls, callus, and suspension cultures. Low amounts of alkaloids were detected in calli even though they were found to retain some laticifer-like cells. Nonembryogenic cell suspension cultures did not contain morphinanes whereas embryogenic cell suspension did contain small amounts of these alkaloids.

Galewsky and Nessler,[47] while studying the synthesis of morphinane alkaloids in *P. somniferum* somatic embryogensis, observed thebaine, the only morphinan positively identified in tissue extracts and in spent growth media. Accumulation of this alkaloid in growth medium parallels its appearance in somatic embryos.

In conclusion undifferentiated cell cultures of *Papaver* species do not produce morphinan alkaloid. On differentiation laticifers are formed; this is probably an important factor for the production of morphinan alkaloids, but other, still unidentified factors also play the role. On the other hand, differentiation is not necessary for the production of sanguinarine.

Nonmorphinan Alkaloids

Furuya and co-workers[83,84] identified a series of benzophenan-thridine alkaloids in *Papaver* cell cultures. Khanna and co-work-ers[66,85] reported the presence of noscapine, narceine, and papaver-ine in poppy cell cultures. Noscapine was also found by Jawadekar et al.[55] Furthermore, protopine,[83,84,88] cryptopine[74,84,87,89] and magnoflorine[83] have been identified in *P. somniferum* cell cultures. Morris and Fowler[90] identified noscapine and narceine as the major alkaloids in *P. somniferum* cell suspension cultures; no morphinan alkaloids could be detected. Orientalidine and isothebaine have also been identified in cell cultures of *P. somniferum*.[91]

Bioconversions

The first reported biotransformation by means of cell cultures of *P. somniferum* was the conversion of thebaine to codeine.[82] As *P. somniferum* cell cultures only produced alkaloids derived from (S)-reticuline (e.g. sanguinarine) but none of the alkaloid derived from (R)-retiuline. Furuya et al.[93] administered (R.S.)-reticuline to the cell suspension cultures. After three days alkaloids were iso-lated. Two alkaloids derived from (S)-reticuline were identified: Cheilanthifoline and scoulerine. A third alkaloid isolated was iden-tified as pure (R)-reticuline. Thebaine, morphine, and codeine were not metabolized by these cell cultures. However, the cells were capable of stereospecifically reducing codeinone to codeine. Tam et al.[94] also found the same bioconversion. Conversion of thebaine to neopine was reported. Codeine, neopine, papaverine, and D.L.-Lau-danosoline were not metabolized. Enzymatic reduction of codei-none to codeine was also achieved with cell free preparations of whole plants of both *P. somniferum* and *P. bracteatum*.[95] Yeoman and co-workers[96,97] reported the use of in reticulate polyurethane immobilized *P. somniferum* cells for this bioconversion.

Cell free extracts of the whole plant were shown to be able to convert (R,S)-reticuline to salutaridine.[98] The enzymatic conver-sion of codeine to morphine in isolated poppy capsules was re-ported by Hsu et al.[99] In addition, cultures of cells in an embryo-genic state were able to convert codeine to morphine at a low rate, although the major metabolic products of codeine were N-oxides.[100]

Verpoorte[64] in conclusion observed that so far production of morphinan alkaloids in large scale cultures of *Papaver* species has been very insignificant. Several approaches to increase production by elicitation and immobilization have failed. A strategy using molecular biology to eventually increase alkaloid production by means of genetic engineering is hampered by the fact that no morphinan alkaloid production occurs in cell culture. Isolation of enzymes to identify the genes responsible for biosynthesis of morphinan alkaloids has yet to be done but this is a difficult task. The vast knowledge which is accumulating on the biosynthesis of other isoquinoline alkaloids[101] may contribute to a better understanding of the pathway leading to morphinan alkaloids. Transformation of *Papaver* species with *Agrobacterium rhizogenes* has been proven to be feasible[102] thus opening the way for genetic engineering.

REFERENCES

1. Hutchinson, J. The Families of Flowering Plants, Vol. I, Reprint Edition. Macmillan and Co., London, 1960.

2. Hocking, G. M. A Dictionary of Terms in Pharmacognosy. C. C. Thomas Publisher, Springfield, Illinois, 1955.

3. Darwin, C. The Effects of Cross and Self Fertilization in Vegetable Kingdom. Murray, London, 1876.

4. Veselovskaya, M. A. The Poppy. Its Classification and Importance as an Oleiferous Crop. Amerind Publishing Co., New Delhi (translated from Russian), 1976.

5. Musalevski, A. and Teodosievski, D. A. Contribution on the Problem of Morphine Contents of the Opium and of the Capsule in Population of Opium Cultivated in Macedonia. Pl. Breed Abst. 1973, p. 43, 4505.

6. Okada, Satoshi, Hiroshi Nakahara, Chikako Yomato, Tadashi Shibata, Kenshu Mochida and Hiroshi Isaka. Morphine Content of Japanese Opium Harvested during 1981-1985. Bull. Nat'l. Inst. Hyg. Sci. (Tokyo), 1986, (104), pp. 163-164.

7. Fedde, F. Papaveraceae, In Die Naturfichen Pflanzenfamilien (ed. Engler, A. and Prantl, K.) 1936, 2 Augl 17B, p. 59.

8. Leger, J. Appareil Vegetatif des Papaveracees. Mem. Soc. Linn Normandi, 1894-1895, p. 18.

9. Sugiura, T. Chromosome Studies in Papaveraceae with Special Reference to the Phylogeny, Cytologia, 1940, 10:558-576.

10. Yasui, K. Cytological Studies in Artificially Raised Interspecific Hybrids of *Papaver*. Cytologia, 1936, p. 7 and 1937, p. 8.

11. Kuzmina, N. E. Cytology of the Cultivated Poppy, in Connection with its Origin and Evolution. (Sum in English.) Bull. Applied Bot. of Gen. and Plant Breeding, 1935, Ser. II, No. 8. Cited by Sugiura.

12. Fedorov, A. A. Chromosome Numbers of Flowering Plants. (Russ) Akad. Nauk SSSR Moscow, 1969, pp. 483-484.

13. Koshy, Jolly K. and Mathew, P. M. Cytological Studies in a Few Species of the Papaveraceae from South India. Cytologia (Tokyo), 1988, 53(4), p. 647.

14. Farmilo, C. G., Rhodes, H. L. J., Hart, H. R. L., and Taylor, H. Detection of Morphine in *Papaver Setigerum* DC. Bull Narc., 5(1) 1953, pp. 26-31.

15. Hammer, K. and Fritsch, R. The Questions of the Ancestral Species of Cultivated Poppy (*P. somniferum* L.). Kulturpflanze, 1977, 25:113.

16. Hirishi, N. J. Cytological Studies in *Papaver somniferum* and *P. setigerum*. Genetica, 1960, 31, p. 1.

17. Sharma, J. R. Discrete Phenological Variation in Opium Poppy (*P. somniferum* L.). Indian Drugs, 1981, 19, p. 70.

18. Gupta, R., Khanna, K. R., Singh, H. G. and Gupta, A. S. Status Report on Opium Poppy. ICAR, New Delhi (Memio), 1978.

19. Roelofs, E. T., Phenotypical and Genotypical Eversporting Varieties. Genetica, 1937, 19, p. 465.

20. Ilieva, S., Krusheva, R. (Krovsheva, R.) and Mateeva, D. The Morphological Nature of Some Modifications of the Flowers of Opium Poppy. Pl. Breed Abstr., 1978, 48, p. 9925.

21. Belyaeva, R. G. and Nevkrytaya, N. V. A Phenological and Genetical Analysis of Mutants with Altered Flower Structure in *Papaver somniferum* L. Pl. Breeding Abstr., 1979, 49, p. 10360.

22. Voskerusa, J. Investigation on Morphine Content in Poppy. Pharmazie, 1960, 15, p. 552.

23. Coiciu, F., Mlesnita, L., Stefanescu, A. and Ragz- Kotilla, E. Comparative Trials with Line and Varieties of Poppy. Ann. Inst. Gene. Agron. Bucuresti C., 1960, 27, p. 163.

24. Rustembekov, S. S. and Argynbaev, T. S. Features of Alkaloid Content in Poppy in Relation to Plant Morphogenesis. Referationyi Zhurnal, 1977, 11, p. 55.

25. Nymann, V. and Hansson, B. Morphine Content Variation in *P. somniferum* L. as Affected by the Presence of Some Isoquinoline Alkaloids. Hereditas, 1978, 88, p. 17.

26. Kaul, B. L., Tandon, V. and Choudary, D. K. Cytogenetic Studies in *Papaver somniferum* L. Proc. Ind. Acad. Sci. Bull., 1975, 88, p. 321.

27. Floria, I. G. and Ghiorghita, G. I. The Influence of the Treatment with Alkylating Agents on *P. somniferum* in M_1. Plant Breed Abstr., 1981, 51 (10), p. 9050.

28. Gohil, R. N. and Kaul, Ranjana. Genetic Stocks of Kashmir (India) Papaver. Nucleus (Calcutta), 1978, 21(3), pp. 219-233.

29. Patra, N. K., Chauhan, S. P. and Srivastava, H. K Syncytes and Premeiotic Mitotic and Cytomictic Comportment in Opium Poppy (*Papaver somniferum*). Indian J. Genet. Plant Breed, 1987, 47(1), pp. 49-54.

30. Khanna, K. R. and Gupta, R. K. An Assessment of Germplasm and Prospects for Exploitation of Heterosis in Opium Poppy (*Papaver somniferum* L.). Contemporary Trends in Plant Sciences (ed. S. C. Verma). Kalyani Publishers, New Delhi, 1981, pp. 368-381.

31. Husain, A. and Sharma, J. R. The Opium Poppy. Central Institute of Medicinal and Aromatic Plants, Lucknow, 1983, pp. 1-167.

32. Sarkany, S., Danos, B. and Sarkany-Kiss. Studies of Heterosis Effects in Poppy Hybrids. Ann. Univ. Sci. Budapest, 1959, 2, p. 211.

33. Michna, M. and Szwadiak, J. Influence of Crossing on Seed Yield and Morphine Contents in F_1 Hybrids of *P. somniferum* L. Nasiennictwo, 1964, 8, p. 579.

34. Singh, U. P. and Khanna, K. R. Heterosis and Combining Ability in Opium Poppy. Indian J. Genet. 1975, 35, pp. 8-12.

35. Kaicker, U. S., Choudhury, B., Singh, B. and Singh, H. P. Breeding of Opium Poppy. P. Breed. Abstr., 1976, 46, p. 9546.

36. Saini, H. C. and Kaicker, U. S. Manifestation of Heterosis in Exotic and Indignous Crosses of Opium Poppy. Indian J. Agric. Sci., 1982, 52, p. 564.

37. Popov, P., Dimitrov, J., Deneva, T., Georgiev, S. and Iliev, L. New Cultivers of Opium Poppy (*P. somniferum*). Rasteniev'd Nauki, 1981, 18(6), pp. 67-72.

38. Anon. NBRI Newsletter XV, 1, Jan. 1988. National Botanical Research Institute, Lucknow.

39. Khanna, K. R. and Sudhir, Shukla. HPLC Investigation of the Inheritance of Major Opium Alkaloids. Planta Medica No. 2, April, 1986, pp. 77-162, 157-158.

40. Singh, U. P. and Khanna, K. R. Male Sterility in Opium Poppy (*P. somniferum*). Sci. and Culture, 1970, 26, p. 556.

41. Khanna, K. R. and Singh, U. P. Genetic Effect on Irradiation on Opium Poppy. Proc. 1st All India Cong. Cytol Genet. Chandigarh, 1971, pp. 312-326.

42. Ilieva, S., Mekhandzhier, A., Mateeva, D. and Dimitrova, S. Biological and Biochemical Changes in *P. somniferum* L. Under the Influence of Ionizing Radiation. Herba Polonica, 1975, 21, p. 412.

43. Furusato, K. Polyploid Plants Produced by Colchicine. (Jap. Summ. English.) J. Bot. and Zool., 1940, 8 (8), pp. 1303-1311.

44. Andreev, V. S. The Increase of Morphine Contents in Polyploids of the Opium Poppy (Papaver Somniferum L.). Proc. Acad. Sci. U.S.S.R., 148, 1963, p. 206.

45. Trease, G. and Evans, W. C. Pharmacognosy, XII Edition, Bailliere Tindall, London, 1983.

46. Ghiorghita, G. I., Elvira, V. Gille and Ecaterina, T. Toth. Gamma Irradiation, Ethylmethanesulfonate and Diethylsulfate Treatments Induced Changes in Morphine Contents and Other Biochemical Parameters in *Papaver somniferum*. Rev. Roum Biol. Ser. Biol. Veg., 27(2), 1982, pp. 121-126.

47. Galewsky, Samuel and Nessler, Craige L. Synthesis of Morphinan Alkaloids from Opium Poppy (*Papaver somniferum*) Somatic Embryogenesis. Plant Sc. (SHANNON), 45(3), 1986, pp. 215-222.

48. Phillipson, J. D., Thomas, O. O., Gray, A. I. and Sariyar, G. Alkaloids from *Papaver armeniacum, P. fugax* and *P. tauricola*. Planta Med. 41, 1981, p. 105.

49. Singh, S. P. and Khanna, K. R. Heterosis in opium poppy (*Papaver somniferum* L.). Indian J. Agric. Sci., 61(4), 1991, pp. 259-263.

50. Ljungdahl, H. 1922. Zur Zytologie der Gattung Papaver. Vorlaufige Mitteilung-Svensk Bot. Tidskr, 16,1: pp. 103-114.

51. Sugiura, T. 1936a. A list of chromosome numbers in angiospermous plants II. Proc. Imp. Acad., Tokyo, 12,5: pp. 144-146.

52. Sugiura, T. 1937a. Studies on the chromosome numbers in higher plants with special reference to cytokinesis II–cytologia, Fuji Jub. Vol pt 2: pp. 845-849.

53. Sugiura, T. 1936b. Studies on the chromosome numbers in higher plants with special reference to cytokenesis 1. Cytologia, 7, 4: pp. 544-595.

54. McNaughton, I. H. 1960. Internal breeding barriers in Papaver. Scottish plant breeding sta. Rept. 1960: pp. 76-84.

55. Sugiura, T. 1938. A list of chromosome numbers in angiospermous plants. V. Proc. Imp. Acad. Tokyo, 14,10: pp. 391-392.

56. Yasui, K. 1937a. Cytogenetic studies in artifically raised interspecific hybrids of Papver. VII P. somniferum LxP bracteatum Lindl. Cytologia, 8, 2: pp. 331-342.

57. Sugiura, T. 1937c. A list of chromosome numbers in angiospermous plants. IV Pro. Imp. Acad. Tokyo, 13, 10: p. 430.

58. Feinbrun, N. 1963. Chromosomes of some East-Mediterranean Papaver species. Caryologia, 16, 3: pp. 649-652.

59. Sugiura, T. 1937b. A list of chromosome numbers in angiospermous plants III Bot Mag. (Tokyo) 51, 606: pp. 425-426.

60. Tischler, G. 1934. Die Bedeutungender polyploidie fur die Verbreitung der Angiospermen, erlautert an den Arten Schleswig-Holsteins mit Ausblicken auf andere Florengebiete Bot. Jahrb. 67: p. 1-36.

61. Sugiura, T. 1931. A list of chromosome numbers in angiospermous plants. Bot. Mag (Tokyo) 45, 535: pp. 353-355.

62. Tahara, M. 1915f. The chromosomes of Papaver. Bot.Mag. (Tokyo) 29, 344: p. 254, 257.

63. Beal, J. M. and Maud, P. F. 1939. The Merton Catalogue. A list of chromosome numbers of species of British flowering plants. New Phytol. 38, 1: pp. 1-31.

64. Vepoorte, Robert, Van Der Heijden, Robert, VanGulik, Walter M., and Hoopen, Hens J. G. Plant biotechnology for the productions of alkaloids: present status and prospects in the Alkaloids. Ed. Arnold Brossi. *40*,81-354, 1991. Academic Press Inc., New York.

65. Tam, W. H. J., Constabel, F. and Kurz, W. E. W. Phytochemistry, 19, p. 486, (1980).

66. Khanna, P. and Khanna, R. Ind. J. Exp Biol. 14, p. 628 (1976).

67. Kama, K. K., Kimoto, A. F., Hsn, A. F., Mahlberg, P. G. and Bills, D. D. Phytochemistry 21,219, 1982.

68. Hodges, C. C. and Rapoport, H. J. Nat Prod. 45,481, 1982.

69. Heinstein, P. F. J. Nat. Prod. 48.1, 1985.

70. Kamimura, S. and Akutsu, M. Agricu Bio. Chem. 40,907, 1976.

71. Kamimura, S. and Nishikawa, M. Agric, Biol. Chem. 40,907, 1976.

72. Kamimura, S., Akutsu, M., and Nishikawa, M. Agric. Biol. Chem. 40,913, 1976.

73. Yoshikawa, T. and Furuya, T. Planta Med. 51,110, 1985.

74. Staba, E. J., Zito, S. and Amin. M. J. Nat Prod. 45,256, 1982.

75. Griffing, L. R., Fowke, L. C., and Constabel, F. J. Plan physiol, 134, 645, 1989.

76. Schuchmann, R. and Wellmann, E. Plant Cell. Rep. 2,88, 1983.

77. Zito, S. W. and Staba, E. J. Planta Med. 45,53, 1982.

78. Nessler, C. L. and Mahlberg, P. G. Can J. Bot, 675, 1979.

79. Cassels, B. K., Breitmaier, E. and Zenk, M. H. Phytochemistry 26, 1005, 1987.

80. Kutchan, T. M., Ayabe, S., Krueger, R. J., Coscia, E. M. and Coscia, C. J. Plant Cell. Rep. 2, 281, 1983.

81. Rush, M. D., Kutchan, T. M., and Coscia, C. J. Plant Cell Rep 4,237, 1985.

82. Czygan, F. C. and Abou-Mandour, A. Dtsch Apoth Zig 126, 1060, 1986.

83. Syono, Aikuta K. and Furuya, T, Phytochemistry 13, 2175, 1974.

84. Furuya, T, Syono K., and Alkula. Phytochemistry, 11, 3041, 1972.

85. Khanna, P., Khanna, R. K., and Sharma, M. Indian J. Exp. Biol. 16,110, 1978.

86. Jawadekar, V. A., Sadi, A. N. and Deshmukh, V. K. Indian J. Pharm Sci, 47, 84, 1985.

87. Forche, E. and Frautz, B. Plants Med. 42, 137, 1981.

88. Lockwood, G. B., Phytochemistry, 20, 1463, 1981.

89. Anderson, L. A., Homeyer, B. C., Phillipson, J. D. and Roberts, M. F. J. Pharm. Pharmacal. 35. 21P, 1983.

90. Morris, P. and Fowler, M. W. Planta, Med 39, 284, 1980.

91. Lockwood, G. B. and Pflanzenphysial, Z. 11,4361, 1984.

92. Grutzmann, K. D. and Schroter, H. B. Abhdl. Akadem wiss (Berlin) Kl. Chemie, 3,347, 1966.

93. Furuya, T., Nakano M. and Yoshikawa, T. Phytochemistry 17,891, 1978.

94. Tam, W. H. J., Kurz, W. G. W., Constabel, F. and Chatson, K.B. Phytochemistry, 21,253, 1982.

95. Hodges, C. C. andRapoport, H. Phytochemistry, 19,1681, 1980.

96. Corchete, P. and Yeomann, M. M. Plant cell Rep. 8,128, 1989.

97. Goy, J. G. and Yeomann, M. M. in "7th International Congres on Plant Tissue and Cell Culture." P. 360 Abstn Amsterdam, 1990.

98. Hodges, C. C. and Rapoport, H. Biochemistry, 16, 3729, 1982.

99. Hsu, A. F., Liu, R. H. and Piotrowski, E. G. Phytochemistry 24,473, 1989.

100. Hsu, A. F. and Rack, J. Phytochemistry. 28, 1879, 1989.

101. Galneder, E. and Zenk, M. H. in "Progress in Plant Cellular and Molecular Biology." (H. J. J. Nijkamp, L. H. W. Vander Plas and J. Van Aartrijk Ed.) p. 754 Kluwer Academic Publishers, Dordrecht. The Netherlands 1990.

102. Williams, R. D., Archambault, J., and Ellis, B. E. in "7th International Congress on Plant Tissue and Cell Culture." p. 356 Abstr. Amsterdam, 1990.

Chapter 4

Agricultural Studies

Papaver somniferum is cultivated in different parts of the world where the soil and climatic conditions are congenial for its healthy growth. The plant is delicate and needs vigilant care in its culture. It can grow in sandy loam or clay loam soil which is well ploughed and pulverized. High organic matter will facilitate both water-holding capacity and drainage capacity to avoid water stagnation.

The poppy in India is grown in almost all kinds of soils, viz. clayey, sandy-loam, loamy sand, and sandy clay, but the plant is preferably grown on the sandy loam type of soil. Such a soil presents a uniform appearance and is fairly moisture retentive and easily cultivable and productive.

Opium poppy is cultivated as a rabi-crop and often follows a crop of maize or other kharif (rainy season), viz. capsicum, or groundnut crop. Since poppy cultivation is still a family affair, the individual holdings under cultivation are small. The same fields are utilized for food crops during the portion of the year when poppy is not cultivated. It is very rare when the fields are allowed to lie fallow before the sowing of poppy seed. Land is prepared in the plains of India in September through October by repeated ploughing and harrowing and brought to a fine tilth; 25-37 cartloads (20-30 tonnes) of farm yard manure are added per hectare four to six weeks before sowing seeds. Penning sheep and goats on the field is preferred. Application of phosphate and nitrogenous fertilizer has a beneficial effect on opium yield.

To facilitate sowing, small seeds of opium poppy are mixed with fine earth or ash and broadcast in October through November at the rate of 3 kg/hectare. The field is then reploughed or harrowed so as to bury seeds uniformly and then divided into squares of 2-3 meters leaving an interval which is raised about 2 to 2 1/2 cm, a channel is then excavated on these ridges to carry water to every square. Frequent light irrigation is necessary until seedlings are fairly well established. When the seedlings are 5-7.5 cm high they are thinned out 20-25 cms apart and weeded.

Gupta[4] advocated that 30 cm apart line sowing of poppy seed produced higher yield than broadcast sowing besides promoting interculture and lancing operations. A plant population of 300,000/ha is conducive to higher yields of opium which is normally 30kg./ha.

Ramanathan et al.[1] reported that in agro-climatic conditions around Jabalpur (India) poppy yielded the highest amount of opium (60.2 kg/ha) when sown on November 15. Yield decreased to 52.4 kg/ha when seeds were sown on November 25 and any further delay in sowing caused rapid reduction of opium yield.

On the other hand, Dabral and Patel[2] observed highest opium yield (50.4 kg/ha) from October 19 sowing followed by November 4 sowing (45.1 kg/ha). Subsequent studies revealed that due to higher temperature during October, germination is hindered in early sowings. On the other hand, late sowings caused fast maturity of the crop, resulting in a smaller-sized capsule having less opium. First half of November, therefore, was suggested to be an optimum period for poppy sowing in India.

Germination takes place some five to 20 days after sowing depending on the moisture condition of soil. Since the poppy does not readily take to transplanting, the young seedlings are thinned out, when they are approximately seven to 14 days old. Poppy seed sown at 3 cm depth in the fields having 75 percent available moisture would bring more yield of opium as well as poppy seeds, making cultivation more economical.[16] A judicious irrigation (1-3 waterings) is desirable at this time. Depending on the weather conditions, irrigation is generally supplied at an interval of ten days. All cultivation is done by hand and the plots are meticulously weeded.

FERTILIZERS AND MANURES

The farmers try to maintain a rich top layer of humus by adding large amounts of farm yard manure (FYM) (10 to 20 ton/ha) to the poppy fields.[3] Gupta[4] recommended 30-50 kg of P_2O_5 and wherever Potash is deficient, K_2O should be applied at 25-30 kg/ha. Both these fertilizers are recommended to be placed 5-6 cms deep in the rows before sowing.

Nitrogen (calcium ammonium nitrate) should be divided into three equal splits, the first dose given at sowing and subsequent doses given 30-40 days and 60-65 days after sowing. Total amount of this fertilizer may vary from 60 kg/ha to 90 kg/ha depending on the amount of FYM added to soil. For Zinc deficiency 12-20 kg/ha of zinc sulphate should be added.[4]

Ramanathan et al.[6] observed highest opium and morphine yields at 75 kg N and 50 kg P_2O_5/ha. P alone at higher doses decreased both opium and morphine yields. Morphine percentage however remained unaffected up to 75 kg P_2O_5/ha but increased further up to 100 kg P_2O_5/ha. Progressive increase in opium yield up to 100 kg each of N, P_2O_5, and K_2O was also reported.[5]

It was reported that nitrogen at the rate of 125 kg/ha and P_2O_5 at 40 kg/ha gave the highest yield of opium. Out of different cake (or extracted refuge made into cakes) used as organic fertilizers, Karanj cakes were better. Among micro-nutrients, cuO_4 @ 250 ppm as foliar spray was beneficial. The Azobactor culture A_2 used in Gulabia variety of *Papaver somniferum* gave maximum yield.

Schrodter[7] reported that higher morphine in the capsule was observed at medium (50-60 kg.ha.) level of N and limited supply of P and K. Whole plant morphine was however increased at fertilizer rates of up to 140 kg N/ha.

It was observed in a status report that N-P-K application in 2:1:1 ratio increased plants, fresh weight and yield of opium, morphine, and seeds. Slight variation in the proportion of N in this mixture caused reduction in all the components except morphine yield.[5]

It has been observed at Udaipur (Rajasthan) that phosphate fertilization up to 40 kg P_2O_5/ha increased the latex, seed, and capsule husk yield while the addition of potash showed no significant increase in yield.[102] In another study it was shown that split application of 90 kg N/ha, 1/4 at sowing, 1/2 at rosette, and 1/4 at preflowering stages gave the highest yield of latex.[103] With response to N it was also observed that some cultivars of *Papaver somniferum*, viz. 1G42 yielded 17.6, 29.8, and 9.8 percent higher latex, seed, and capsule husk, respectively, as compared to Dholia Dwarf. With regard to nitrogen, application of 90 kg N/ha gave 41.2, 37.5, and

58.0 percent higher yield of latex, seed, and capsule husk over 30 kg N/ha, respectively.[104]

Kharwara et al.[106] observed that in sandy loam soils at Palampur the crop sown on November 14 and 24 significantly increased opium and seed yield and also improved morphine in opium and oil in seed than the crop sown on December 14 and 24. The authors also observed that application of N at 150 kg/ha produced significantly higher opium and seed yield and morphine in opium than the lower dose of 75 kg/ha.

Yadav et al.[8] reported that nitrogen application (50, 100, 150, and 200 kg/ha) increased the opium, seed, and morphine yields as compared with the control. Morphine concentration (percent) in the opium, however, increased up to 100 kg/ha and decreased when N doses exceeded that level. Divided application of N, i.e., half at sowing and the remainder at the stem elongation stage, proved beneficial for opium, seed, and morphine yields.

Kharwara et al.[9] reported a significant increase in the opium and seed by applying 150 kg N + 100 kg P_2O_5/ha. Similarly, morphine content of opium and oil content of seed increased with nitrogen. Higher dose of phosphorus decreased the morphine and protein contents but increased the oil contents. Nitrogen in three equal splits, at sowing, flowering, and capsule initiation, was better than single or double application.

Turkhede et al.,[10] however, observed that application of N and P increased opium, morphine, and seed yield but improvement in morphine percentage was obtained with N application. Maximum increase in the three characters was recorded with 50 kg N and 10.75 kg P/ha.

Opium yield, morphine yield, and morphine percentage remained unaffected by plant densities. Seed yield was influenced only when fewer plants per unit area gave higher seed yields.

Cheema et al.[11] observed that seed germination percentage and seedling vigor of 'Dhoha' opium poppy improved in response to seed treatment with Dithane M-45 and Dithane Flowable (Mancozeb). Complete inhibition in germination was observed when the treated seeds were stored for one year at ambient conditions. Gupta[4] advised that treating 1 kg seed with 4-5 g Thiram or Brassicol (Quintozene) before sowing protects it from soil-borne diseases.

Twenty kg/h of 10 percent BHC or 5 percent Aldrin should be applied to the soil at the time of land preparation to protect the poppy seedlings against termite and shoot-cutting caterpillars.

Foliar spray of urea on opium poppy gave significant results.[12] Spraying of 3 percent wt/vol solution of urea at intervals of four days gave the highest yields of opium (63.93 kg/ha) and morphine (5.58 kg/ha) per unit area.

Spraying of one of the micronutrient products "Tracel" or "Micron special" 75-90 days after sowing is recommended for commercial cultivation.[13] The yield of opium was observed at 60 kg/ha and above in most of the treatments.

Earlier Ramanathan[14] reported the increase of opium, its morphine content,[36] and morphine yield per unit area when micronutrients like B, Cu, Fe, Mg, Mn, Mo, and Zn alone and in combination are applied to opium poppy. Laughlin[15] emphasized the role of boron in poppy cultivation. He observed that hand-drilled B at 2 kg/ha in boron-deficient alluvial soils of Australia increased capsule plus seed yield by 700 percent while sprays of B (2 kg/ha) caused only 500 percent yield increment.

Influence of foliar application of Triacontanol (Tria) on opium poppy was studied under glass house conditions. Plant height, capsule number and weight, morphine content, CO_2 exchange rate, total chlorophyll, and fresh and dry weight of the shoot were significantly increased at 0.01 mg/L Tria. At the highest concentrations (4 mg/L) total chlorophyll, CO_2 exchange rate, and plant height were significantly inhibited. Thebaine and codeine contents remained unaffected at all concentrations. The concentrations of Fe, Mn, and Cu in shoots were maximum at 0.01 and Zn at 0.1 mg/l Tria. Increase in shoot weight, leaf area ratio, and chlorophyll content were significantly correlated with morphine content.

IRRIGATION

The poppy crop needs an assured supply of irrigation water. Ten to 12 irrigations in light soil and seven to eight irrigations in heavier soils are generally advised.[4] Lack of moisture from the rosette stage until the appearance of the capsule causes heavy loss in latex yield. Irrigations are, however, withheld during the lancing period in India when latex is collected from the capsules. Gupta et al.[5] recommended ten to 14 irrigations at an interval of ten days but Raman-than et al.[1] suggested a 20-day interval.

It has been observed that irrigated plants contained more alkaloids than unirrigated ones.[16] Moisture stress during main growth period and flowering had an adverse effect on morphine accumulation.[17] Melke,[18] however, reported highest alkaloid content in plants grown on muck soil at lower moisture level. Lyakin[19] indicated that rosette is a highly sensitive stage for moisture availability in soil and suggested that 21-23 percent moisture should be maintained in soil at the rosette stage of the plant. It was however observed that prolonged moisture stress from rosette formation to flowering caused reduction in opium and morphine yield.[17]

WEEDING AND HOEING

Germination takes place some 5 to 20 days after sowing depending on moisture content of the soil. Within a week or so the first four leaves appear and two or three weeks later the stem begins to elongate. The plant reaches full development in about two months depending on the agroclimatic conditions.[20]

The initial growth of the plant is very slow. The first weeding and hoeing is recommended three to four weeks after sowing. A second weeding should be carried out 55 days after sowing. These operations (weeding and hoeing) are aimed at simultaneously thinning the seedlings to keep a distance of 10 cm between plants in rows. To maintain the plant population to the recommended level, one or two more weedings/thinnings are done until the formation of the flowering shoot.[4]

Application of Asulex (Asulam Methyl 4-amino-phenyl-sulpho-nyl carbomate) (6 liters/ha) or Dicuron (chlortoluron) (1.5 kg/ha) before seedling emergence controls most weeds in light soils and chlorcoluron (1.5 kg/ha) in heavy soil. Handweeding has, however, been found to be ideal.[4]

Foldesi and Bernath[21] used Diuron [3-(3,4-dichlorophenyl)-1] @ 1.5 kg/ha very effectively for the control of weeds in Hungary. Mezotox (nitrofen) @ 8-10 kg/ha was used to eliminate thinning in mechanized cultivation.

For postemergence application 2 kg Diuron 1-dimethylurea/ha at 8-10 leaf stage was recommended.[16] Reglone (Diguate dibromide) was applied as postemergence to prevent weeds like *Sinapsis, Raphanus,* and *Amaranthus.* For weeds belonging to Cruciferae and Compositae families Diuron + Plakin (Asulam) @ 2 + 8 kg/ha, respectively, was used as postemergence after application of Diuron + Reglone. For control of weeds of Chenopodiaceae and Gramineae a combination of Diuron + Dual (metolachlor) at a rate of 2 + 1 kg/ha, respectively, proved most effective. This combination controled *Chenopodium album,* an obnoxious weed in poppy cultivation. Due to the synergistic effect of both herbicides, this combination even killed weeds 20 cms. tall.[16]

Singh et al.[22] applied Sencor (Metsibuzin), Linuron (Tsifluralin) and Patoran (Metobromuron), at 1/2 and 1/4 kg/ha and observed that these chemicals reduced weed population significantly but also killed the poppy crop completely. Alachlor (alachlor "Lesso") and Nitrofen (2.4-dichlorophenyl 4-nitrophenyl ether) also acted similarly. When poppy seeds were kept in between two layers of farm yard manure (FYM=mixed livestock manure, mostly rotten), Alachlor and Nitrofen did not damage the crop but controlled monocot weeds effectively.

Other chemicals remained toxic to the crop even when FYM was used. The FYM proved a useful protectant to the sensitive crop of poppy. High microbial population developed on FYM band perhaps decomposed herbicide particles and made the seed row free from herbicide injury and the inter-row spacing free from weed population.[23]

HARVESTING

Opium is harvested by incising (lancing) the capsules at a particular phase of plant growth known as industrial maturity, when the capsules become fully swollen and green, but are still immature. This stage comes 15-21 days after the petalfall. The period of collection of opium extends from the end of March to the end of April in the plains of India, but in the hills it is extended until June.

Some 90-100 days after sowing seed (or 75-80 days after germination) the plants, which are now waist-high, begin to flower.

Usually after three days the petals have fallen, and after an additional ten to 14 days the capsules are ready for lancing[24] (Figures 4.1-4.3).

The lancing operation is performed by skilled workers. The field is usually divided into three portions so that each portion gets a chance for scarification of the capsule every third day. When 5-10 percent of capsules in a portion of the field are ready the lancing, operation commences. The capsules are incised with a special type of knife "Nastar," a home-made gadget comprised of three or four tines or small blades with sharpened points tied together with thread in such a way as to create a space of about 1-2mm between the blades. All these blades are fixed to a holder some 18 cm long. The depth of the incision is controlled by the affixation of the tines of the holder. The "nashta" is held carefully as one would hold a pencil while writing and the incision is made. This Nashtar ensures uniformity of the depth (0.3-0.4 cms) of the incisions. Usually each capsule is lanced three or four times and sometimes as many as five or six times until no more latex exudes. The incisions are usually made vertically from below upward or by downward swift strokes starting below the stigmatic rays. In India lancing is done after

FIGURE 4.1. A section of poppy field showing flowers and fruits in Barabanki (Lucknow). (Photo by L. D. Kapoor)

FIGURE 4.2. Lancing of poppy capsules in poppy field. (Photo by L. D. Kapoor)

mid-day, i.e., during the hottest part of the day. The reason for this is that a *"pellicle"* is said to form on the surface of the freshly exuded latex as a result of the hot sun.[24] This may sound easy but in fact is an art.

The depth of the incision is controlled by the affixation of tines or blades to the holder, if the incision is too deep the latex is exuded to the interior of the capsule and is thus lost. If the cut is too shallow the yield of latex will be low. The laticiferous vessels are mostly located between the epicarp and mesocarp and run from below upwards in the capsule.[20] An experienced skilled worker knows the optimum depth of lancing and 150-200 capsules can be lanced in an

FIGURE 4.3. Lancing operation–A closer view. (Photo by L. D. Kapoor)

hour.[24] The duration for harvesting latex or raw opium lasts for one to two weeks after which the capsule dries up and is harvested for seed. Gupta[4] reported that on bright sunny days between 12 noon and 4 p.m. the capsule is given a longitudinal incision by a special knife (naka) which has three or four thin sharp-edged blades fitted at a distance of 1.5 to 2 mm from each other; it produces an equal number of incisions to a depth of 0.4 cm. Each capsule is incised four times at an interval of three or four days to obtain most of its latex. The latex continues to exude from the cut end and is deposited over its surface. This is scraped the next day before 10 a.m. and collected in earthen or plastic containers. The raw opium is churned every day to make it homogenous and then water content is decanted off. The narcotic department makes payment to the grower for opium of 70 percent consistency as a standard unit (Figures 4.4 and 4.5).

The lanced crop is left in the field for the next 20-25 days so that its seeds get fully matured. These drying capsules are picked, spread in open yards for further drying, and thrashed to obtain the seed crop.

YIELD OF OPIUM/MORPHINE

The yield of opium varies from place to place depending on various factors, such as a variety or cultivar of *P. somniferum* L., nature of the soil, sowing time, climate and weather, the number and mode of lancing of capsules, etc. Each capsule gives a maximum yield of latex at the first lancing and the yield decreases with each successive lancing. Use of a knife having six blades instead of three or four may increase the total yield of opium per capsule. Terminal capsules yield more opium of considerably higher morphine content than the later ones.

On an average, the current yield of opium in India is between 25 and 30 kg per acre. As the cultivation of opium is only permitted under license, a cultivator's license may not be renewed if this production falls below the average yield.[25] The Turkish method of spiral incision yielded higher opium than the Indian method of vertical incision.

The average seed yield of poppy is about 50 kg/ha. A bold dry capsule of poppy weighs about 7 g and contains 11,000-12,000 seeds weighing about 3.5 to 4 g.[16]

Ramanathan[26] observed the highest latex in early lancing from March 26 and continued with ten days' interval. The highest amount of morphine was collected at 98 days after germination.

Kleinschmidt and Mothes[27] observed highest alkaloid concentration in 12-week old seedlings.

Pfeifer and Heydenreich[28] reported that 15-20 days after flower-

FIGURE 4.4. Farmers deliver raw opium in earthenware to the government weighing center against cash payment. (Photo by L. D. Kapoor)

FIGURE 4.5. Weighing of raw opium at the government weighing center. (Photo by L. D. Kapoor)

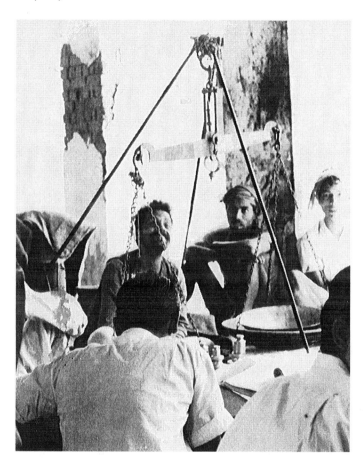

ing, morphine content reaches its maximum though the secondary alkaloids have not reached their highest concentration. Neubauer[29] observed varietal differences with regard to distribution of morphine, codeine, thebaine, and narcotine for three stages of capsule maturity and also observed a rather constant morphine content after rosette stage. Sarkany et al.[30] observed that the relative contents of the total alkaloids is lesser in the main capsule in comparison to the younger side capsules.

Ramanathan and Ramchandran[1] and Ramanathan[14] reported highest morphine in opium of first lancing out of three or four lancings made on each capsule. It has been reported that after eight days of capsule formation there is no decrease in morphine content with ripening. Maximum opium yield per capsule was obtained in 16 to 17-day-old capsule. It declined afterward.[5]

Khanna and Shukla[31] studied the effect of capsule position and number of lancings numbers on content of major alkaloids in *Papaver somniferum*. They reported that morphine content declined in successive lancings and from top (main capsule) to the lower (side) capsules. The maximum codeine content (4.4 percent) is reached in the side capsule during the second lancing. Similarly, thebaine content was found to be maximum during second lancing in the side capsule. The behavior of narcotine was irregular. Some of the highest values were observed in the lowest capsule during the third lancing. Papaverine could be observed by them only during the first lancing of the main capsule.

Although morphine and other alkaloids, viz., codeine, thebaine, narcotine, and papaverine, are derived from the latex after incision of capsule at biological maturity, the other vegetative parts of opium poppy, viz., leaf, root, stem and flower, do contain these alkaloids at various stages of growth of the plant.

It was reported that the traditional method of lancing of capsules gave higher yield of alkaloids but the seed and capsule-shell were more in unlanced capsules. The percentages of five major alkaloids increased with growth of capsule, but lancing reduced their contents. However, codeine and thebaine contents remained nearly unchanged.[107]

Kapoor[32] reported the presence of opium alkaloid in seedlings when their first foliage leaves expand, after sowing, and when they are 2.5 cm long and about three weeks old. The sepals, petals, ovary, and capsule all contain the opium alkaloids, but the stamen and seed are devoid of any alkaloids. But Kapoor has not identified which alkaloids were observed.

Laughlin[101] harvested poppies at weekly intervals beginning ten days after bloom and continuing until four weeks after the dry commercial harvest stage. At each harvest, the plants were cut off at ground level and partitioned into terminal capsules, lateral capsules, seed, and the combined stem plus leaf components. The morphine concentration of both terminal and lateral capsules reached a maximum value of 1.1 percent six weeks after full bloom and then decreased by about 10 percent at the dry harvest stage. The morphine concentration of stems and leaves also reached a maximum of 0.1 percent about the same time as capsules but decreased rapidly and had halved that concentration at dry commercial harvest. The maturity-compensating factors of decreasing morphine concentration gave similar total plant morphine yield at any time of harvest from two to seven weeks after full bloom. The morphine extracted from the whole plant at these times of harvest was about 50 percent greater than that derived from capsules alone at the time of dry commercial harvest.

In Villar-1, maximum contents of morphine, codeine, and narcotine were found in capsules at biological maturity (0.595, 0.132, and 0.020 percent, respectively) and thebaine and papaverine (0.780 and 0.082 percent, respectively) at the preceding technical maturity. The total alkaloid content at biological maturity was 0.826 percent.

In Novinka-198, maximum content of all five alkaloids (0.8858 percent) was recorded in capsules at biological maturity. At blooming stage, percent distribution of morphine in different plant parts was: whole plant 0.09-0.18; leaves 0.06-0.07; stem 0.01-0.03; capsule 0.04-0.230; and stalks 0.07-0.19. Morphine content of young roots was about 0.39 percent and of old about 0.13 percent.[16]

It is reported that young plants had the highest morphine contents in their roots. But at blooming stage, the upper one-third portion of the stem had the maximum morphine.[16] During capsule formation, morphine starts accumulating there.[34] Poethka and Arnold[35] also observed highest morphine in roots at seedling stage. With growing, age of plant morphine content of leaves increased and that of roots decreased. During capsule formation, leaves translocated their morphine to the capsule. Highest morphine content was observed in the capsule either at ripe or at half ripe stage. Declining trend in alkaloid content of root and increasing trend of that of leaf and stalk with the age of crop was also observed.[28] Poppy head showed two maxima by a distinct minimum. The first occurred in 14-day-old fruit and the other after full maturity. Nikonov[36] also observed 70 percent of the total alkaloids in capsules. He further reported that in seedling stage, root contains maximum alkaloids, from where these are translocated to stem and other parts of the plant. It has also been reported that the capsule is the main depository of alkaloids. The alkaloids are absent in seed.[32]

CHANGES IN MORPHINE CONCENTRATION

Heeger and Schroder[37] and Miran and Pfeifer[38] reported a progressive increase in morphine concentration right up to dry maturity, but Bunting[39] and Schroder[40] found morphine concentration at a maximum when the capsules were green or semidry and it gradually declined toward dry harvest maturity. Rain had the leaching effect on morphine concentration of capsules after dry maturity. Kopp[41] reported significant reductions in morphine if harvested capsules were merely stored in a moist environment. Miczulska[42] has shown a close association between the extent of fungal colonization in the capsule and the decline in morphine concentration over this period.

It was observed that the dry substance production of the capsule wall and placenta reaches maximum in the "glassy" stage of the seed buds (sixth or seventh day after flowering). The mass of the dry substance of the seed increases until the seeds begin to color, up to the twenty-fourth or twenty-fifth day after flowering. At the same time, the alkaloid accumulation differs in different parts of the capsule. The total alkaloid level of the placenta reaches its maximum (14, 9°C) before the seeds begin to color. However, the accumulation continues and even increases (up to 23, 1°C) in the capsule wall for an additional four to eight days. During the intensive water loss of the capsule these maxima decrease by 20 to 40 percent.[109]

DISEASES AND CONTROL MEASURES

Opium poppy is susceptible to many fungal and insect infections. To ensure the economic harvest from the capsules adequate plant protection measures are necessary.

Downy Mildew

Downy mildew is one of the serious diseases caused by *Peronospora arborescens* (Berk) de Bary and was observed epiphytotic in many parts of the world. Fokin[43] reported that *P. arborescens* caused hypertrophy and curvature of the stem and flower stalks. The infection spread upward from the lower leaves.[44] The entire leaf surface was found to be covered by a downy mildew coating composed of conidiopores and conidia of the pathogen. The stem, branches, and even capsules are attacked resulting in premature death of the plant. It was found to cause extensive damage to the crop cultivated on moist sites.[45]

The disease appears annually on the crop from seedling stage to maturity in the opium-growing areas in India. The formation of the capsule is adversely affected; hence, the opium yield is significantly reduced.[16]

Darpouse[46] recommended seed disinfection and spraying of seed beds with 0.5 percent Bordeaux mixture and other copper fungicides for control of disease.[47] Alavi[48] observed that the disease is seed transmitted and recommended three sprays of Bisdithane at 0.15 percent followed by Benlate (Benomyl) at 0.05 percent for effective control of the disease. Crop rotation and fungicidal sprays with Gramisan (Terbutryn), Germisan (Carboxin + maneb), or dusting with Thiram (Tetramethyl thiuram disulphate), Gramisan, Germisan, or Agrosan (Phenylmercury acetate) were suggested as useful protective measures against the disease.[49]

Rathore et al.[50] observed that for control of secondary infection of downy mildew caused by *Peronospora arborescens* spray of Metalaxyl (Ridomil 25 wp) (menthyl N-(2-methoxyacetyl) N-(2,6-xylyl)-DL-alaminate) at 0.2 percent at 15 days, interval resulted in maximum increase in latex and seed yields.

Powdery Mildew

Another disease, powdery mildew, caused by *Erysiphae polygoni* did severe damage to the crop in Rajasthan (India).[51] Another powdery mildew caused by *Oidiopsis* sp. has been observed in India and France.[52,46] One spray of Spersul (sulphur) (0.5 percent) for the control of disease caused by *E. polygoni*[51] and seed disinfection for the control of powdery mildew caused by *Oidium* sp. has been recommended.[46]

Seed-Borne Diseases

There are few reports on seed-borne diseases on opium poppy. *Fusarium scirpi* var. *caudatum* was observed as a semiparasite capable of attacking capsules and seeds in Bulgaria.[53] *Pleospora calvescens* was detected in seeds in Denmark and Poland.[54,55] *Pleospora papaveracea* and *Peronospora orborescens* were also found associated with poppy seeds in Turkey.[56] Seeds from Denmark were found contaminated with *Dendryphion penicillatum*.[57] Samples collected from different localities in Poland indicated that *Helminthosporium papaveris* was constantly associated with the seeds. Although no mycelium was observed in embryo or endosperm, in seedling stage the fungus infected the primary cortex.[58]

Seedling Blight

Angell[59] found that *Pythium ultimum* and *P. mamillatum* were responsible for the seedling blight, a serious disease in Australia. Experimental studies revealed that seedling blight was physiogenic as well as pathogenic, well distributed throughout Australia. Fungicidal treatment of seed failed to control this disease in sterilized soil.[60]

Leaf Blight

Leaf blight disease of opium poppy caused by Helminthosporium species was reported to be one of the most destructive diseases in Bulgaria, Switzerland, Czechoslovakia, the Netherlands, Romania, and Yugoslavia.[16]

It was reported that *Pleospora calvescens* (syn. *P. papaveracea*) penetrated the foliar tissue and ramified through the inter-cellular spaces.[61] During the course of pathogenesis toxins were released by the parasite enabling it to assimilate nutrients from the host tissue. Higher rainfall and relative humidity in June or July favor the infection.[62] In Romania, the high plant density and poor nutrition favor infection by *Helminthosporium papaveris*.

Some control measures have been suggested. Early sowing was observed to reduce infection.[63,49] Seed disinfection or spraying of seed beds with 0.5 percent Bordeaux mixture or any other copper fungicides were also suggested for the control of the disease.[46,47]

Matzner[64] observed that incorporation of lime as $CaCo_3$ at 285 kg/ha appreciably reduced the losses caused by *H. papaveris*. Fungicidal sprays with Systox (Demeton) before flowering reduced infection and increased the yield.[65] Seed disinfection with Organol (maneb + metiram) at 0.5 and 1 kg/100 kg significantly reduced the infection caused by *H. papaveris*.[66] Misko[67] showed that soil treatment with borate and manganese superphosphate (30 kg/ha) reduced the infection significantly. However, spraying with Germisan (1 percent), Gramisan (0.1 percent), and dusting with Thiram, Gramisan, Germisan, or Agrosan were found to be the most effective control measures of the disease.[68] It is also reported that seed dressing with Zaprawa (Carboxin + theram) and Nasienna-T (Nasimant hydrolysed proteins) and spraying plants with Synkotox and Dithane M-45 (Mancozels) were also most effective prophylactic measures against *H. papaveris*.[49]

Kishore et al.[69] recorded three fungal diseases of opium poppy in Eastern Uttar Pradesh, namely (1) damping off opium seedlings (*Fusarium solavi* (Mart) Sacc); (2) leaf spot (*Alternaria alternata* (Fr) Keisler); and (3) leaf blight (*Drechslera spicifera* (Bainier) V. Arx state of *Cochliobolus spicifer* Nelson).

Capsule Infection

Macrosporium papaveris caused large velvety black spots on green capsules.[70] *Alternaria brassicae* var. *somniferum* also infected the capsules.[71] Miczulska[72] observed that increased infection caused by *Alternaria alternata* significantly reduced the morphine content. Felklova[73] similarly reported decrease in codeine and thebaine contents in the infected capsule. Capsule rot of opium poppy caused by *Dendryphion penicillatum* was prevalent in Rajasthan.[74] A serious capsule rot caused by *Botrytis cinerea* was observed in Lucknow.[16]

Capsule infection can be controlled by periodic spraying of fungicides. Four to six fungicidal applications using 200-500 g/ha of Bavistin (Carbendazim) will be useful to control most of fungal phytopathogens. This will also be a prophylactic measure for maintaining a healthy crop.[16]

Wilt and Root Rot

Root rot caused by *Macrophomina phaseoli* was reported from Rajasthan.[75] The symptoms of disease were withering and drying of leaves, early maturity of capsule, and low opium yield. A fusarium root rot caused by *F. semitectum* inflicting serious damage especially in seedling stage has been recently reported in Uttar Pradesh.[16]

Viral Diseases

Extensive damage to opium crops by several viral diseases has been reported. Cabbage ring spot virus infected opium crops[76] and the infection soon became systemic.[77] Another suspected virus disease causing yellowing of plants and elongation of stem was observed in the former Czechoslovakia.[78] The occurrence of bean yellow mosaic virus on poppy was reported in Bulgaria. Symptoms appeared as irregular chlorotic bands along the veins which spread very fast all over the interveinal areas, leaving green stripes along the veins. Infected plants were mostly stunted.[79] Turnip mosaic virus infected opium poppy in Hungary.[80] A turnip mosaic virus was identified in the former Czechoslovakia as a causative agent of the mild form of disease which was manifested by deformation of leaves with symptoms of green-yellow mosaic, by flower breaking and by weak development of poppy heads with the symptoms of mosaic and stripes. Strong attack caused dwarfing, distinct mosaic, necrosis, and deformation of leaves, stalks, and heads, and the plant's decline.[105] Turkoglu[81] reported the occurrence of mosaic disease in Turkey. Infected plants showed stunting, chlorosis, and distortion of leaves, as well as reduced seed and opium production. Muller et al.[82] found the occurrence of mycoplasma-like organisms in the phloem tissue of infected poppy plants which showed dwarfing and witches'-broom symptoms. Only mosaic disease of opium poppy was observed to date in India, causing stunting, vein banding, and deformed capsule formation.[83] The virus could be transmitted readily from sap. Thermal inactivation of the virus was

50-55°C and had only limited hostage. The disease was transmitted by aphids as well as sap.

Another pest *Orobanche papavaris* (Broom rape),[84] a root parasite, does considerable damage to the poppy crop in India. It is rampant in districts of Manasa and Sawan of Madhya Pradesh and considerably reduce the opium yield. Swabbing of 0.3 percent solution of allyl alcohol and 1 ml of teapol carries practically 100 percent mortality in 72 hours of *Orobanche papaveris*.

WEATHER HAZARDS

High and gusty winds during the opium season are dilatorious because they dry up the plant and thus check the exudation of latex.[85] Hail storms ruin the crop while a heavy rainfall, especially during the period of lancing and collection of latex, leaves little or nothing to collect.

Sunburning of leaves is another disaster in which the leaves get dried up and wither with more or less discolored purply-black or brownish veins, the pith decaying from above downward. Plants exhibit these symptoms both in poor and rich soils when the weather is hot and there is deficiency of moisture in the soil. Under these conditions water update by the roots fails to keep pace with the leaf transpiration. Frost is also very destructive sometimes. During heavy frost the thermal balance of the protoplasm of the cells is lost. The protoplasm in such cases shrinks and the cells die. The only remedy appears to be to water the field by the morning after the frost when the plants will again try to regain their proper balance and the crop may be saved.

Other Pests

In India, rats, rabbits, monkeys, blue bulls, and parrots also do considerable damage to the poppy crop.[85]

Apart from fungal and viral diseases, opium plant is attacked by aphids, thrips, and cutworms.

Opium cutworm (*Agrotis suffusa* Hubn) does great harm to the young opium poppy. Cutworms remain burrowed in the soil during the day and eat away the leaves during the night; the affected plants die after a few days.

They are controlled by flooding the fields with water. The cut-worms float on the surface and are picked up by birds. Caterpillars do serious damage to the growing crop. Irrigating the fields dislodges them from their soil haunts, and then they are eaten by crows and myna. *Heliothis armigera* Hubn is another opium pest which develops while the plant is young and subsists at first on its tender leaves. As the plant matures, it eats its way up the stem and finally bores into and eats the interior of the capsules. Small insects and crickets (*Gryllotalpa vulgaris* Latr.) are reported to cause some damage to the crop. Poppy seeds for sowing are stored in vessels containing camphor to prevent them from insect attack.[84]

INSECT PESTS

Hussain and Sharma[16] have compiled an exhaustive list of insect pests right from seedling stage to harvest stage. Major pests of opium poppy are briefly reported below.

    I. Pests associated with root damage.

        a. Root Weevil (*Stenocarus fuliginosus* Marsh)

            This insect pest bores into upper parts of roots which turn black and leaves become brownish and dull.[86] The larva unusually mines the leaf-lamina also. Chemical control measures, biological control, and change in agricultural practices gave good results. Complete control of larvae was obtained by application of 12 percent BHC dust at 2-3 kg/ha together with seed and superphosphate at the time of sowing.[87]

    II. Pests associated with leaf-stem damage.

        a. Aphids (*Aphis* sp., *Myzus* sp., *Rhopalosiphum* sp., *Acyrthosiphon* sp.)

            Aphids suck the sap from the plant and transmit the virus to the poppy.[88] The mosaic virus in Hungary caused serious damage to the poppy crop.[89] The Plum pox virus, necrotic intermediate, and yellow strains of virus have been reported to do considerable damage to opium poppy in France.[90] Aphids can be controlled by treatment of the crop with systemic insecticides at the time of attack.[88] Spray of 0.08-0.10 percent Prim-carb (Flumetralin) 800-900 L/ha gave reliable protection from *A. fabae* Scop on opium poppy.[91] Spray with metosystox or malathian gave satisfactory aphid control.[92]

III. Pests associated with floral damage.

a. Thrips (*Frankliniella* spp.)

Thrips have been found to attack opium poppy in Formosa, Madras, and Bara Banki (Lucknow). These are transferred from one host plant to another by bees moving in the opium poppy field. No chemical control has been standardized to control the thrips. Removal of petals within an hour or two of blooming is considered the best cultural control.[93]

b. Sawfly (*Corynis similis* Mocs)

The caged larvae of *Corynis similis* Mocs. (*Tenthredo dahli*) were found feeding on opium poppy.[94] They feed first on pollen and petals and later on ovules and capsules. The bionomics, extent of damage, and its control measures are not precisely known.

IV. Pests associated with capsule damage.

a. Head Gallfly (*Clinodiplosis* sp.)

The larvae of this insect develop beneath the petals, the capsule is deformed, and the development of seed is checked. Later the capsule is attacked by fungi and the capsule dies. This pest was first observed in Sweden in 1944 and in the former Czechoslovakia in 1959. There are no records of its control measures.

b. Capsule Weevil (*Ceulorrhynchus mascula-alba* Hbst.)

The female lays the eggs in capsules where the grub develops and comes out after two to three weeks for pupation. The climatic conditions of Central Europe are considered very favorable for the outbreak of this weevil.[95] In Romania the relative value of morphine/codeine increased in the dried capsules which were attacked by poppy weevil (*Stenocarus fuliginosus* or *C. denticulatus*).

The weevils appear at the time of blossoming and feed on leaves, flowers, and stem. One capsule is attacked by one or two grubs. The capsules are deformed, seeds are mostly destroyed, and those which ripen have a moldy taste. Sometimes seed germination takes place inside the capsule.[97] Dusting with 2.0 percent parathion methyl (dimethyl-O-4-nitrophenyl phosphorothioate methyl parathion) at 25 kg/ha or 12 percent B.H.C. at 30 kg/ha was found to be a very effective treatment against this pest.[87]

c. Capsule Borer (*Heliothis* spp.)

This insect attacks in highest intensity the capsule of opium poppy where the cultivated area is low, has more moisture content in the soil, and where gram crop is adjacent to field of opium poppy. The larvae of this pest first feed on leaf but later on the seed of opium poppy, inserting the larval head in the capsule and living there until the emergence of the moth from pupa. Due to this infection, the seed yield and production of latex are reduced. This infection has been recorded in Bara Banki (Lucknow)[98] and in the former Soviet Union.[99] For chemical control Pandey[100] recommended the quinalphos 25EC (O,O-diethyl O-quinoxalin-2yl phosphorothioate) at 150 ml/ha and carbaryl W.P. (1-naphthyl methycarbomate) at 0.75 kg/ha in the form of 0.025 and 0.125 percent concentrations, respectively.

REFERENCES

1. Ramanathan, V. S. and Ramchandran, C. Opium Poppy Cultivation, Collection of Opium, Improvement and Utilization for Medicinal Purposes. In "Cultivation and Utilization of Medicinal and Aromatic Plants." RRL Jammu Tawi, 1977, pp. 38-74.

2. Dabral, K. C. and Patel, O. P. Poppy Cultivation in Chhindwara District. Preliminary Studies on Opium Seed and Capsule Yield. J.N.K.V.V. Res. Jour., 1975, 9, p. 73.

3. Anonymous. CIMAP Newsletter, 1981, 8, no. 3. CIMAP (CSIR) Lucknow.

4. Gupta, Rajendra. Improved Production Methods for Opium Poppy. Indian Hort., 1984, Jan.-March, p. 9.

5. Gupta, R., Khanna, K. R., Singh, H. G. and Gupta, A. S. 1978, Status Report on Opium Poppy. ICAR, New Delhi (Memo).

6. Ramanathan, V. S., Ramchandran, C. and Prakash, V. Effect of Organic Manure and Chemical Fertilizers on Opium, Morphine and Poppy Seeds. Indian Jour. Agron., 1975, 20 (1), p. 65.

7. Schrodter, H. Untersuchugen uber Veranderungen des Morphingehalts reifender Mohnkapseln. Pharmazie, 1965, 20, p. 169.

8. Yadav, R. L., Mohan, R., Singh, R. and Verma, R. K. The Effect of Application of Nitrogen Fertilizers on the Growth of Opium Poppy (*Papaver somniferum*) in North Central India. J. Agric. Sci., 102(2), 1984, pp. 361-366.

9. Kharwara, P. C., Awasthe, O. P. and Singh, C. M. Effect of Nitrogen, Phosphorus and Time of Nitrogen Application on Yield and Quality of Opium Poppy (*Papaver somniferum* L.). Indian J. Agron, 1986, 31(1), p. 26.

10. Turkhede, B. B., Rajat, De, Ramanathan, V. A. and Sewa, Ram. Effect of Nitrogen and Potassium Rates and Plant Densities on the Opium, Morphine and Seed Yield of Opium Poppy (*Papaver somniferum*). Indian J. Agric. Sci., 1981, 51(9), pp. 659-662.

11. Cheema, H. S., Chakravarti, B. P. and Thakorelal, B. B. Effect of Opium Poppy Seed Treatments with Dithiocarbamates on Germination, Host Vigour and Viability of Seeds in Storage. Indian J. Plant Pathol., 1987, 5(2), pp. 180-183.

12. Ramanathan, V. S. Effect of Foliar Spraying Urea on the Yield of Opium and Its Morphine Content in Opium Poppy (*Papaver somniferum*). Indian J. Anim. Res., 1982, 16(1), pp. 23-28.

13. Ramanathan, V. S., Upadhyay, and Surana, J. L. Effect of Foliar Spraying Micronutrient Products on *Papaver Somniferum* on the Yield of Opium and Morphine. Indian J. Agric. Res., 1984, 17(3), pp. 132-136.

14. Ramanathan, V. S. Effect of Micronutrients on the Yield of Opium and Its Morphine Contents in Opium Poppy. Indian J. Agric. Res., 1979, 13, p. 85.

15. Laughlin, J. C. The Boron Nutrition of Poppies (*Papaver somniferum* L.) on Chernozem and Alluvial Soils of Tasmania. Planta Med., 1979, 36(3), p. 245.

16. Hussain, A. and Sharma, J. R. The Opium Poppy. Central Institute of Medicinal and Aromatic Plants, Lucknow, 1983.

17. Turkhede, B. B., De, R. and Singh, R. K. Consumptive Water Use by Opium Poppy. Indian J. Agric. Sci., 1981, 51(2), p. 102.

18. Melke, J. Effect of Soil and Moisture Levels on the Content and Accumulation of Alkaloids in Opium Poppy (*Papaver somniferum*) and Deadly Nightshade (*Atropa belladonna*). Annales universitatis mariae curie-sklodovska C. 1978, 33, p. 55.

19. Lyakin, A. S. The Effect of Precipitation on Poppy Growth, Development and Yield. Vopr. Agronomii Frunze Kirgis SSR, 1977, pp. 157-160.

20. Anonymous. The Opium Poppy. Bull. Narcotics, 1953, July-Sept.

21. Foldesi, D. and Bernath, J. Effective Method for Control of Tolerant Weeds in Poppy Cultivation (*Papaver somniferum* L.). Planta Med., 1979, 36(3), p. 248.

22. Singh, R., Turkhede, B. B. and Singh, R. K. Use of Farm Yard Manure as a Protectant in Chemical Weed Control in Opium Poppy. Indian J. Agron., 1980, 25(2), p. 308.

23. Klingman, G. C. Weed Control as a Science. John Wiley and Sons, New York, 1961, pp. 64-80.

24. Krikorian, A. D. and Liedbetter, M. C. Some Observations on the Cultivation of Opium Poppy (*Papaver Somniferum* L.) for its Latex. Botanical Review., 41(f), 1975. pp. 30-47.

25. Anonymous. Report on the Operations of the Narcotic Department for the Year Ending September 30, 1962. New Delhi.

26. Ramanathan, V. S. Study on the Deteriation of Morphine and its Preservation by Chemicals in the Fresh Latex of Opium Poppy (*Papaver somniferum* L.). Part II. Indian J. Agric. Res., 1980, 14, p. 6.

27. Kleinschmidth, G. and Mothes, K. Investigations about Biogens is of the Alkaloids of *Papaver somniferum*. Arch. Pharm. Ber. Dentrsch Ges., 1960, 293(10), p. 948.

28. Pfeifer, S. and Heydenreich, K. Die Akkumulation der Mohnalkaloide Zwischen Blute and Biologischer Reife. Ein Beitrag Zum Proglem Gewinnug Von Alkaloiden aus Grummohn. Pharmazie, 1962, 17, pp. 107-114.

29. Neubauer, D. Planta Med., 1964, 12, pp. 43-50.

30. Sarkany, S., Sarkany-Kiss, I., and Verzar Petri, G. Abh. Deut. Akad. Wiss. Berlin. K.I. Chem., 1966, 3, pp. 355-362.

31. Khanna, K. R. and Sudhir, Shukla. HPLC Investigation of the Effect of Capsule Position and Lancing Number on Content of Major Alkaloids in *Papaver Somniferum*. Planta Medica, 1986, No. 2, April, pp. 161-162.

32. Kapoor, L. D. The Laticiferous Vessels of *Papaver somniferum* L. PhD Thesis, London University, London, 1958.

33. Ramanthan, V. S. A Study on the Deterioration of Morphine and Its Preservation by Chemicals in the Fresh Latex of Opium Poppy. Part I. Indian J. Agric. Res., 1979, 13(4), p. 229.

34. Wegner, E. Die Morphinvert lung in der Mophnpflanze und ihre Veranderung in lange der vegetationsperiode als Beitrag Zur physiologie diseases. Pharmazie, 1951, 6, p. 42.

35. Poethke, W. and Arnold, E. Untersuchungen uber den Morhingehalt der Mohnpflanze. Pharmazie, 1951, 6, p. 406.

36. Nikonov, G. K. Accumulation of the Main Alkaloids in the Opium Poppy in the Course of Its Ontogenesis. United Nations Bulletin on Narcotics, 1958, 10(1), p. 70.

37. Heeger, E. F. and Schroder, H. Untersuchungen uber die Morphinertrage bei *Papaver somniferum* L. unter mitteldeulschen Anbauverhaltnissen. Pharmazie, 1959, 14, p. 228.

38. Miran, R. and Pfeifer, S. Uber die Veranderumgon Alkolsidhaushalt der Mohntanze Wahrend einer vegetations periode. Scientia Phermaceutica, 1959, 27, p. 34.

39. Bunting, E. S. Changes in the Capsule of *Papaver somniferum* between Flowering and Maturity. Ann. Appl. Biol., 1963, 51, p. 459.

40. Schroder, H. Untersuchungen uber Veranderungen de Morphingehalts reifendez Mohnkapseln. Pharmazei, 1965, 20, p. 169.

41. Kopp, E. Versuche zur zuchtung einer morphin reiohen mohnsorte. Pharmazie, 1957, 12, p. 614.

42. Miczulska, I. Observation on the Influence of Infection of Poppy (*P. somniferum* L.) by Parasitic Fungi on the Morphine Content in the Poppy Head. Roczn. Nauk. Roln. Ser. A. Rosb., 1967, 93, p. 189.

43. Fokin, A. D. Diseases and Injuries of Cultivated Plants Observed During the Summer of 1922 in the Government of Vyatka. Tran. Fourth All Russian Entomo. Phytopath. Congress. Moscow, 8-14 Dec. 1922, p. 108.

44. Yossifovitch, M. Kingdom of the Serbs, Croats and Solvenes. A Destructive Disease of the Poppy in Southern Serbia. Internat. Bull. Plant Protection, 1928, II, p. 18.

45. Pietkiewicz, T. A. From Studies on Diseases of Oleagenous Plants. Roczn. Nunk. Rol., 1958, 78, p. 199.

46. Darpouse, H. A Contribution to the Study of the Diseases of Oleagenous Plants in France. Ann. Epiphyl. N.S., 1945, XI, p. 71.

47. Beaumont, A. Diseases of Poppies. Gdners. Chron. Ser., 1953, 3, p. 71.

48. Alavi, A. How to Prevent and Control Downy Mildew of Opium Poppy. Iranian J. Plant Path., 1975, p. 73.

49. Rudny, R. The Effect of Sowing Time, Methods of Cultivation and Protective Measures on the Health and Yield of Opium Poppy (*Papaver somniferum* L.). Variety Mak niebieski KM. Prace Naukowe Instytutu Ochrony Roslin, 1976, 18, p. 167.

50. Rathore, R. S., Mathur, Sneh and Mather, Kusum. Control of Secondary Infection of Opium Downy Mildew Induced by *Peronospora arborescens* by Metalaxyl. Summa Phytopathol, 1986, 12, pp. 202-206.

51. Kothari, K. L. and Prasad, N. Powdery Mildew of Opium Poppy in Rajasthan and Its Control. Indian Phytopath., 1972, 25, p. 36.

52. Kothari, K. L. Other Powdery Mildew on Opium Poppy (*Papaver somniferum* L.). Indian Phytopath., 1968, 21(X) p. 456.

53. Christoff, A. Some Plant Diseases New to Bulgaria: 2nd Contribution. Bull Soc. Bot. de Bulgarie, 1934, VI, p. 37.

54. Neergaard, P. (Thirteenth Annual Report from the J. E. Ohlsen Phytopathological Laboratory. 1st August 1947 to 31st July 1948.) 1949, p. 19.

55. Zarzycka, H. Fungus Flora of Opium Seeds. Roczn. Nauk. rol., 1958, 78 (Ser. A), p. 309.

56. Gobelez, H. Research Work on the Varieties and Areas of Spread of Bacterial and Parasitic Diseases Affecting and Contaminating the Seeds of Cultivated Plants Growing in Certain Provinces of Central Anatolia as well as the Approximate Degree of Damage Caused by Such Diseases. Zir. Fak. Yayinl., 1956, 107, pp. 62, 131.

57. Neergaard, P. "Sixth and Seventh Annual Reports Relating to the Control of Seed Pathalogy from 1st April 1953 to 31st May 1954 and 1st June 1954 to 31st May 1955." 1956, p. 17.

58. Blotnicka, K. Harmfulness of Helminthosporium papaveris, Sawada (Perfect state *Pleospora calvescens* (Fries) (Tulasane), for Opium Cultivation). Hodowla Roslin, Aklimatyzacjai Nasiennietwo, 1976, 20, p. 59.

59. Angell, H. R. Seedling Blight II. Soil in Relation to Seedling Blight of Opium Poppy and Peas. Aust J. Agric. Res., 1950, 1, p. 132.

60. Angell, H. R. and Hills, K. L. Seedling Blight III. Control of Seedling Blight of Opium Poppy by Liming. J. Aust Inst. Agric. Sci., 1951, 17, p. 17.

61. Zogg, H. Contributions to the Knowledge of Plant Defence Reactions: The Influence of Temperature on the Development of the Gummous Demarcation Zone. Ber. Schweis. Bot. Ges., 1945, VI, p. 507.

62. Bogarada, A. P., Lyman, V. E. and Taranich, A. P. On the Resistance of Poppy to Helminthosporiosis. Selekts Semenov., 1971, 36, p. 78.

63. Radulescu, E. and Perseca, E. (On the biology of *P. Papaveraceae.*) Rev. Roum. Biol. Ser. Bot., 1964, 9, p. 19.

64. Matzner, F. The Influence of Fertilizing with Chlorides and Sulphates in Conjugation with Calcium on the Yield and Seed Quality of Poppy (*Papaver somniferum*) in Relation to Infection by Helminthosporium Papaveris (Henning). Wiss Z. Friedrich Schiller Univ., 1958, 7, p. 295.

65. Mass Geesteranus, H. P. A Premature Ripeness of Seed Poppies: Pleospora Papaveracea on Seed Poppies. Tijdschr. Pl. Ziekt, 1960, 66, p. 107.

66. Mraz, F. Study of the Effectiveness of Treatment Compounds on the Conidia of the Fungus Helminthosporium papaveris (Hennig). Ann. Acad. Techecosl. Agric., 1963, 36, p. 51.

67. Misko, L. A. Heminthosporiosis of Poppy. Zashch Rast Moskva, 1963, 8, p. 56.

68. Radulescue, E., Perseca, E. and Docea, E. Contributions to the Control of Helminthosporiosis of the Poppies (P. calvescens) by Agrotechnical and Chemical Methods. Lucr. Stiint. Inst. Agron. N. Balcescu, 1961, Ser B, 5, p. 333.

69. Kishore, Raj, Tripathi, R. D., Johri, J. K. and Shukla, D. S. Some New Fungal Diseases of Opium Poppy (*Papaver somniferum*), Indian J. Plant Pathol., 1985, 3(2), pp. 213-217.

70. Parisi, Rosa. Notes on Some Parasites of Medicinal and Aromatic Plants. Bull. Orto. Bot. Napoli., 1921, VI, p. 285.

71. Grummer, G. The Influence of Alternaria Infection of Poppy Capsule on Promptitute of Germination by Their Seed. Flora Jena, 1953, 140, p. 298.

72. Miczulska, L. Observations on the Influence of Infection of Poppy (*P. somniferum* L.) by Parasitic Fungi on the Morphine Contents in the Poppy Heads. Roczn. Nauk. Roln. Ser. A. Rosl., 1967, 93, p. 189.

73. Felklova, M. Pathophysiological Study of Some Diseases of Medicinal Plants. Referatiunyi Zhurnal Biologiya, 1977, 7G, p. 406.

74. Sehgal, S. P., Gupta, P. J. and Agarwala, J. M. Capsule Rot of Opium Poppy (*Papaver Somniferum*). Rajasthan Journal of Agric. Sci., 1971, 2, p. 61.

75. Deshpande, A. L., Agarwal, J. P. and Mathur, B. N. Rhizoctonia Bataicola Causing Root Rot of Opium in Rajasthan. Indian Phytopath., 1969, 22, p. 510.

76. Dyer, R. A. Botanical Surveys and Control of Plant Diseases. Fing. S. Afr., 1949, 24, p. 119.

77. McClean, A. P. D. and Cowin, S. M. Diseases of Crucifers and Other Plants Caused by Cabbage Ring Spot Virus. Sci. Bull. Dep. Agric. S. Afric., 1952, 332, p. 30.

78. Rozsypal, J. Observations on the Virus of Opium Poppy. Ces. Biol., 1957, 6, p. 438.

79. Kovacnevsky, J. C. H. and Kovachevsky, I. C. "Bean Yellow Mosaic Virus in Bulgaria," Phytopath. Z., 1968, 61, p. 41.

80. Horvath, J. and Besada, W. H. Opium Poppy (*P. somniferum* L.): A New Natural Host of Turnip Mosaic Virus in Hungary. Z. pflanzenschvtz., 1975, 82(3) p. 162.

81. Turkoglu, T. Mosaic Virus of Opium Poppy in Turkey. J. Turk. Phytopath., 1979, 8, p. 77.

82. Muller, H. M., Sorguceva, N. A., Fedotina, V. L., Schmidt, H. B., Kleinhempel, H., Procenko, A. E. and Spaal, D. Investigations on the U.S.S.R. and GDR on the Electron Microscopic Detection of Mycoplasma like Organisms in Plants. Archiv fur phyto pathologic und Pflanzenschutz.,1974, 10, p. 15.

83. Anand, G. P. S. and Summanwar, A. S. Studies on a Mosaic Disease of Opium Poppy. Indian Phytopath., 1981, 34, p. 262.

84. Anonymous. Wealth of India. Raw Materials. Vol. VII, Council of Scientific and Industrial Research, New Delhi, 1965, pp. 231-248.

85. Asthana, S. N. The Cultivation of Opium Poppy in India. Bull. Narcotics., 1954, 6, p. 1.

86. Ranninger, R. Coeliodes Fuliginosus, a Coleopteron Injurious to the Poppy in Australia. Mthly. Bull. Agric. Intell. and Pl. Dis, Rome, 1917, VIII, (7), p. 1068.

87. Bogarada, A. P. Results of Using B. H. C. and Polychlorpinene against Poppy Weevil. Himija. Sel., Hoz., 1967, 5(6) p. 24.

88. Cleij, G. Beet Yellow in Poppy Euphytica. 1961, 10, p. 225.

89. Horvath, J. and Besada, W. H. Opium Poppy (*Papaver Somniferum* L.). A New Natural Host of Turnip Mosaic Virus in Hungary. Z. Pflanzenschvtz., 1975, 82(3), p. 162.

90. Sutic, D. Herbaceous Hosts of Plum Pox Virus in the Papaveraceae. Comtes Rendus des Scances de 1 'Academie d'. Agriculture de France, 1977, 63(6), p. 440.

91. Nagy, F. and Csucs, M. Results of Pest Control Experiments on Poppy (*Papaver somniferum* L.) with Special Regards to Aphids (Aphis Fabae Scop). Herba Hung. 1976, 15(2), p. 45.

92. Pirone, Pascal. Diseases and Pests of Ornamental Plants. John Wiley and Sons, New York, 1943, p. 392.

93. Okuni, T. Insects Injurious to Poppy in Formosa II. Bull. Agric. Exp. Sta. Govt. of Formosa, 1921, No. 142.

94. Scheibelreiter, G. K. Contributions to the Knowledge of Poppy Sawfly, *Corynis similis* (*Mocsary*) (Hymenoptera Cimbicidae). Zeitschrift fur Angewandte Entomologie, 1979, 87(4), p. 393.

95. Schrodter, H. and Nolte, H. W. Field Observations on the Effect of Temperature on the Egg Laying and Larval Development of Opium Poppy Capsule Weevil (ceutorrhynchus macula-alba). Nachr. dtsch. Pflsch. Dienst. Berlin, 1952, 6, p. 67.

96. Greathead, D. J. and Scheibelreiter, G. K. Investigations on the Fauna of Papaver spp. and Cannabis Sativa. Commonwealth Inst. Biol. Control, European Station, Delemont, Switzerland, 1978, p. 39.

97. Kotte, W. Injuries Caused by the Poppy Capsule Weevil. Ceulorrhynchus Macula-alba. Zeitschr. Pflanzenkr., 1948, 55(3-4), p. 81.

98. Singh, Dwijendra and Tripathi, A. K. Insect Pests Complex of Opium Poppy (*Papaver Somniferum* L.) in India (unpublished), 1981.

99. Ostrovskii, N. I. and Drozdovskaya, L. S. The Basic Pests and Diseases of Poppies. Zashchita Rastenii, 1970, 15(11), p. 27.

100. Pandey, S. N. Aphim per langane wale kit evam unke niyantran. Bull. Aphim utpadan Takniki Prasicchan, 1980, p. 17.

101. Laughlin, J. C. The Effect of Time of Harvest on the Yield Components of Poppies (*Papaver somniferum*). J. Agric. Sci., 95, 1980, p. 667.

102. Jain, P. M. Effect of Phosphorus and Potassium on Yield of Opium Poppy. Indian J. Agron., 35(3), 1980, pp. 238-239.

103. Jain, P. M. Effect of Split Application of Nitrogen on Opium Poppy. Indian J. Agron., 35(3), 1990, pp. 240-242.

104. Jain, P. M., Gaur, B. L. and Gupta, P. C. Response of Opium Poppy Varieties to Nitrogen. Indian J. Agron., 35(3), 1990, pp. 243-245.

105. Spak, J. and Koubelkova, D. Occurence of Turnip Mosaic Virus in Opium Poppy (*Papaver somniferum*) in Czechoslovakia. SB UVTIZ (USTAV Vedeckotech inf zemed) ochr Rostl, 26(4), 1990, pp. 257-261.

106. Kharwara, P. C., Awasthi, O. P. and Sing, C. M. Effect of Sowing Dates, Nitrogen and Phosphorus Levels on Yield and Quality of Opium Poppy. Indian J. Agron., 33(2), 1988, pp. 159-163.

107. Srivastava, V. K., Pareek, S. K., Gupta, R. and Maheshwari, M. L. Yield and Alkaloids Profile in Poppy Capsule. Indian J. Pharm. Sci., 51(4),1989, pp. 133-136.

108. Srivastava, N. K. and Sharma, Srikant. Effect of Triacontanol on Photosynthesis, Alkaloid Content, and Growth in Opium Poppy (*Papaver somniferum* L.). Plant Growth Regul, 9(1), 1990, pp. 65-72.

109. Bernath, J. Variation of Dry Substance and Alkaloid Accumulation in the Developing Capsule of Poppy (*Papaver somniferum* L.). Herba Hung, 28(3), 1989, pp. 15-20.

110. Worthing, Charles R. The Pesticide Manual. A World Compendium. 6th Edition 1979. Published by The Bertesh Crop Protection Council, Glasshouse Crop Research Institute, UK.

111. Kidd, Harmish, Harlley, Douglas. Pesticide index. 1988. Royal Society of Chemistry (information services). Printed by Unwin Brothers, LAD, Old Woking, Surrey, UK.

Chapter 5

Physiological Studies

Physiological studies on different phases of growth and development in poppy plant (*Papaver somniferum*) have been undertaken by various workers. These contributions give a better understanding of the development of the plant and its active principles.

Two phases of the life-cycle of the plant where growth is particularly active have been studied, namely the development of the flower into capsule (which incidentally is the important part of the plant medicinally) and development of the seed, seed germination, and development of the seedling.

The small mature seed appears to have enough phytochrome to stimulate germination, as brief red and far red light treatments had no effect on its germination.[1] While studying the temperature requirement for germination of seeds, Volovich and Griff[2] observed that poppy seeds could germinate even at 1°C. They can withstand frost well and even repeated freezing and thawing does not lower germination appreciably. Seedlings and young plants can withstand temperatures as low as –3°C or –4°C.[3] The seeds would easily germinate at temperatures from 8°C to 33°C in darkness but the maximum germination was found at temperature range from 13°C to 33°C.[1]

Anderson and Olssen[4] exposed poppy plants at seedling stage and at three-leaf stage to temperatures of –6°C to –8°C and found that 46 percent were undamaged at the seedling stage while 21 percent were undamaged at the three-leaf stage.

Poppy seeds stored for six years in closed metal boxes at temperature of 20-29°C in summer and 13-23°C in winter showed 74 percent germination.[5] Ultrasonics hastened and increased germination of the seed.[6] Ohashi[7] reported that vernalization at relatively high temperatures results in increased contents of oil, fat and alkaloids. It was reported also that seeds held at 6°C for 20 days with 30-50 percent water content showed early development of the plant and increased opium yield.[8] Earlier it was reported by Hirosi[9] that seed vernalization at 2-6° increased the alkaloid contents of *P. somniferum* by 26 percent. Lecat[10] reported that poppy seeds held for 35 days at 2-3°C and 3 percent humidity gave plants having heavier seed but 15 percent less morphine. Seeds held for eight days at 30°C in solution of $CaCl_2$ and KCl gave plants flowering two to three days later with smaller seeds but 15-25 percent more morphine.

When soaked in eosin, poppy seeds showed on germination a shorter hypocotyle and a longer root. Since the roots often showed negative geotropism, eosin may inhibit the effectiveness of the active auxin.[11]

It was observed that gamma radiation stimulated germination and percentage of germination increased from 62 to 85 percent.[12] Ghiorghita et al.[13] reported that successive treatments of seeds with gamma rays and alkylating agents in the first generation and gamma rays in the second generation induced a high variability of plants and there were isolated individuals of *P. somniferum* with high morphine contents in the capsule (0.53-0.54 percent dry wt.).

Irradiated with CO^{60} the poppy seeds yielded four types of mutations, viz. dwarfs, flower color changes, sterile flowers, and biochemical changes (primarily increased morphine content).[14] Fujita et al.[15] reported that the germinating power of seeds was completely destroyed by vapor heating at pressures of three lb. for two minutes or one lb. for three minutes. The effect of gamma ray radiation was significant at doses greater than 80,000r.

Regarding the occurrence of organic acid and mineral substances in different organs of opium poppy, Coic et al.[28] reported that oxalic acid was most abundant in the seeds. They were rich in magnesium and calcium also.

The cotyledons contain fairly good amount of oil, for which it is commercially cultivated in Europe and Russia. Fairbairn and Kapoor[17] reported that seeds do not contain alkaloids.

The seed germination percentage and seedling vigor of some cultivars of *P. somniferum* improved in response to seed treatment with Dithane M-45 and Dithane Flowable (Mancozeb, 1,-2-ethane-diylbis). Complete inhibition in germination was observed when the treated seeds were stored for one year at ambient conditions.[18] Gorgiev and Dragana[19] observed that treatment of poppy seedlings with chlorocholine chloride seems to be the best method to obtain more alkaloids.

Shikimate dehydrogenase (EC1.1.1.25) was extracted and partially purified from three-day-old germinating plants of *P. somniferum* L. cv Dubsky by Smogrovicova et al.[20] The highest specific activity of the enzyme occurred in roots of three to five day-old germinating plants. Most enzyme activity (89 percent) is present in the fraction of soluble proteins. By polyacrylamide gel electrophoresis, four types of shikimate dehydrogenases were found in crude extracts of dry seed. Embryo, endosperm, root, hypocotyle, and cotyledons contained only three multienzyme forms of shikimate dehydrogenase.

Poppy seedlings showed no external injury to soil application of amitrol but catalase activity was lowered by 60 percent after the first day, respiration gradually decreased, and after the fourth day polyphenoloxidase activity was stimulated.[21]

The seeds of poppy contain oil ranging from 44-56 percent and is rich in unsaturated fatty acids particularly oleic acid. The composition of poppy oil was reported by Veselovskaya.[27]

Glycerides of solid acids (Palmitic and stearic) . . 6.2 percent

Glycerides of oleic acid . . . . 68.0 percent

Glycerides of linoleic acid . . . . 24.7 percent

Unsaponified residues . . . . 0.5 percent

The iodine number . . . . 133-140 percent

The values vary according to geographical locations. Coic et al.[28] reported oxalic acid most commonly found in the seeds which were also rich in calcium and magnesium.

Luthera et al.[29] observed that palmitic, oleic, and linoleic were the major fatty acids at all stages of seed development but there was clear predominance of linoleic acid. The proportion of linoleic acid increased tremendously with the deposition of triacylglycerols and was negatively correlated with linoleic acid. The changes in the fatty acid make-up were quite marked in the early developmental stages than at the later stages of the seed development. The active period of their synthesis lies between 15 and 20 days after flowering. Marin et al.[30] reported that members of Papaveraceae show a higher triacylglycerol but lower free fatty acid content than members of Fumariaceae.

The plant develops a flowering stalk during the budding stage and bears a single floral bud at its tip. The flowering stalk remains bent during the budding stage but becomes straight when the flower opens.

Kohji et al.[22,23] reported that the stalk of *Papaver rhoes* first shows a positive geotropic curvature of "nodding" followed by negative geotropic response by growing upright, and then the flower blooms. Curvature occurs because of enhanced elongation of cells of the convex side and the straightening due to elongation of cells on the concave side of stalk. It was deduced that nodding of the flower stalk at early stages, after flower bud formation, was initiated by the weight of the flower bud. Sugawara[24] demonstrated that application of growth substances to the peduncle of *P. somniferum* causes the normal upright peduncle to bend at the site of application due to enlargement of parenchyma cells and the formation of merismatic tissue.

The plant takes about 100 days to reach flowering stage from the time of sowing but the petals fall off in three to five days after flowering. Benada[25] observed that the ovary of *P. somniferum* has a lower redox potential than the petals. The petals of *P. somniferum* may be white or crimson or red but bear no nectaries. Free[26] reported that the flowers of poppy are vigorously worked out by bees and other insects.

PHOTOPERIODIC STUDIES

Gentner et al.[31] observed that opium poppy is a long-day plant with critical day length for flowering of 14-16 hours. Flowering may be induced by two or more long photoperiods or by a single period of light longer than 24 hours. Flowering stems always lengthen but they sometimes lengthen in absence of flowering, i.e., with application of gibberellic acid. It has been observed that flowering was not controlled by brief red irradiation, far red irradiation, or both. Thus, the action of phytochrome was not shown but its presence was not excluded. Conclusively light seems to control poppy flowering through a so-called high energy reaction.

Mika,[32] in an exhastive photoperiodic study of *Papaver somniferum,* reported that capsules collected 98 days after seed germination contained the highest concentration of morphine and those collected after 98-114 days contained the greatest amount of morphine. He also reported that the number of nodes and dry weight of capsules of plants placed on inductive 18-hour photoperiods 14, 30, 45, or 64 days after seed germination were directly proportional. The two subspecies studied differed in their relative response to varying photoperiods.

On short day lengths the rosette phase is longer, stalk formation to budding is reduced, and the period from budding to flowering is increased.[33]

Plants of *P. somniferum* kept on long-day conditions in the field were potted and changed to short day. The half treated with gibberellic acid on the apex flowered 15 days before the untreated controls.[34]

Bernath and Tetenyi[82] observed that long-day conditions accelerated growth and development of poppy plant at optimal light intensity with pronounced effect on size and weight of capsule. Accumulation of alkaloid was more intensive with increase of light intensity.

Geographic distribution (with respect to latitude and altitude) of the crop also influences alkaloid formation, as the morphine and codeine contents of opium produced in the northern part of Korea were higher than that in the southern part.[16]

This is possibly due to northern and southern climatypes of opium poppy. The area of cultivation is rather sharply differentiated: It is about the 40th parallel of northern latitude which forms the limits of the profitable growing of the above mentioned climatypes.

P. somniferum, especially tetraploids, showed advanced plant development and flowering when grown under 24-hour days while eight and 12-hour days retarded development.[35]

Morphine formation was inhibited with shading and some already present disappeared. Shaded plants produced less fat but the percentage of oil in the seeds is approximately equal to that of the controls.[36]

Ravenna[37] reported that opium poppy injured in May and June had higher morphine contents than uninjured plants.

With regard to studies on transpiration Prokofiev and Kats[38] reported that the rate of transpiration was greater in fruits and inflorescences of poppy than in leaves and was maximal during flowering.

Photosynthesis of the pods, for ten to 12 days after flowering, is equal to that of leaves of the middle whorl and the amount of chlorophyll is approximately that of leaves. If the fruit or the leaves are placed in dark, there is a decrease in the amount of seeds in the fruit.[39]

When the capsules, but not the leaves, were kept covered after flowering, fewer seeds were produced containing 3 percent less oil. When the leaves were covered and the capsules left uncovered, the number of seed was also reduced and their oil content declined from 49-21 percent.[40]

A study of photosynthetic rate and stages of organogenesis in *P. somniferum* indicated that the seasonal dynamics of photosynthesis depends on the time required to pass through the stages of organogenesis.[41]

Moisture content of the ovary is high at the early stages of seed development, then decreases independently of the moisture of the surrounding tissues or of the relative humidity.[38] Rates of lipase activity and fat accumulation were low in young seeds but they increased rapidly during development and then decreased during ripening.[40]

Zoschke[42] reported that the alkaloid synthesis was highly dependent on N nutrition, the morphine content varying from 0.09 to 0.70 percent with various additions of ammonium nitrate.

The salinity of soil has various influence on the amounts and distribution of some macroelements in different organs and in different stages of development, but yield of dry substance is lower and osmotic pressure becomes higher as compared with the control samples.[43]

The physiological maturity of the capsules occurs about 15 days after flowering when it is considered fit for the collection of opium with maximum morphine content. The content of codeine and thebaine are reported to be high after flowering.[44]

It is reported that the decapitated plant accumulated much more thebaine, as the transformation of thebaine to codeine and morphine was inhibited, which could only take place in the developing capsule.[45] Bunting[46] reported that maximum volume and fresh weight of capsule are reached within three weeks of flowering, whereas percent water content remains high for about six weeks. In dry season maximum morphine content and minimum water content occur at the same time, whereas in wet season 50 percent of the morphine accumulated is lost before minimum water content is reached.

Peroxidase activity in the capsule reached a maximum about two weeks after flowering and then declined to zero. Morphine content increased at the same time and then gradually fell except for the variety *soproni,* which remained constant.[83]

RESPIRATION STUDIES

While studying the respiration in developing poppy seed Johri and Maheshwari[47] reported that the excised ovules of *P. somniferum* showed an increased oxygen uptake during or preceding pollination, division of endosperm nuclei, cell wall formation in the endosperm, and elongation of cotyledons. In another study, they observed four peaks of oxygen uptake, viz. (1) at fertilization, (2) free nuclei formation, (3) cell wall formation in the endosperm, and (4) period of rapid cell division in the embryo.[48]

It was observed that during seed development in opium poppy[49] the fructose and glucose are present up to free nuclear stage of the endosperm but decrease when it turns cellular and the sucrose becomes abundant. The N accumulates in two phases, the first coinciding with development of the endosperm and the second with development of the embryo.

Balatkova and Tupy[50] reported that sucrose solutions of uracil and 5-bromouracil increased seed set 70 percent when injected into the ovaries three hours prior to pollination. Morphine content of green young capsules is higher than in mature dried ones. However, storage decreases the morphine content of green capsules but not of mature capsules.[51]

EFFECT OF GROWTH REGULATORS

The growth regulators injected into capsules of *P. somniferum* one to two days after flowering resulted in smaller seed size following benzyladenine, kinetin, and 1AA treatment, but larger following GA or 6-methyluracyl. Most treatments lowered morphine and codeine but some increased thebaine, papaverine, and narcotine.[52] Bernath and Vagujfalvi[53] demonstrated that resistance to unfavorable condition prevailing at the time of ripening was increased by cycocel, a growth retardant. It also increased codeine and thebaine contents. But another chemical dimethyl-sulfoxide which effects membrane permeability decreased the codeine and thebaine contents.

Zhdanova and Rusova[54] sprayed maleic hydrazide on the plant during the period of reproductive growth. The treatment decreased the yield of seed as a result of embryogenesis suppression, but when the treatment was carried out during the second part of the maturation period, the dry mass of seed increased. Ramanathan[55] reported that ethrel (2-chloro ethyl phosphoric acid) sprayed at 200 ppm on 75 and 90 days after sowing enhanced the yield of opium up to 63.3 kg/ha. However, spraying foliage with a combination of maleic hydrazide (2 parts/1,000) and potassium sulphate (1 part/100) increased morphine production without deforming the capsule.[56]

Treatment with kinetin and benzyladenine after flowering produced a narrow capsule; benzimidazole inhibited capsule growth. Benzyladenine also produced larger seeds.[57] Gibberellin did not affect the weight of the capsule but caused further development of the integument of the capsule with reduction of the weight of seed and morphine content.[58]

The aqueous extracts of poppy heads contained certain compounds which changed the polarographic stage of morphine to give a higher morphine content.[59] Studies in Germany showed that morphine content is a hereditary factor that depends on soil and climate for optimum production.[60]

Zdenda[61] observed that high rainfall and low temperature toward the end of the vegetative period of *P. somniferum* lowers the alkaloid content. But in another study, morphine content of plant parts increased with irrigation: 0.411 percent of capsules compared to 0.338 percent in non-irrigated.[62]

Malinina and Ivanova[81] reported a direct relationship between alkaloid accumulation and temperature and inverse relationship between alkaloid accumulation and amount of rainfall. Kuzminska[74] reported that morphine content increased when soil moisture was reduced to 30 percent of field capacity.

Annett,[63] in his poineering work on Indian opium poppy, observed that terminal capsules produce opium of higher morphine content then do lateral. Age of capsule has little effect on percentage of morphine provided it is more than eight days from flowering. Percentage of morphine is not affected by weather or fertilizers but they do affect the yield of opium.

Annet et al.[64] further observed that yield of opium depends more on the size of capsule and the vigor of plant than the method of lancing, number of incisions, length of incision, and the time of day. Heeger and Schroder[65] reported that morphine content was highest ten to 30 days after flowering under Central German cultivation conditions.

Holloway[66] observed that superficial wax on the leaf is the dominant factor governing water repellency. Ramanathan et al.[67] observed that spraying of aqueous solutions of chelating agents alone (a mixture of citric acid, tartaric acid, ascorbic acid, and sodium tripolyphosphate) and with sodium chloride on the surface of the capsule just before lancing and immediately after lancing on the latex oozing gave the best results for preservation of morphine in opium poppy.

Felklova and Vaverkova,[68] while studying the yield of codeine from *P. somniferum* under different climatic conditions, reported that higher content of codeine was found in plants cultivated in warm and dry places. The content of codeine on the area is connected with the amount of created biomass.

EXTERNAL FACTORS AFFECTING OPIUM
AND MORPHINE CONTENT

Mika[32] reported that on a whole-plant basis the greatest concentration of morphine was found in plants harvested 98 days after germination, whereas the greatest amount of morphine was found in plants collected after 98 and 114 days.

Application of nitrogen affects the production of opium and its alkaloids. Various workers, viz. Sheberstov et al.,[69] Nowacki et al.,[70] and Costes et al.[71] reported that alkaloid content increased by nitrogen application, whereas Kinoshite et al.[72] observed that by increasing the nitrogen input, though the yield of opium and of seeds increased, the morphine content remained unchanged. Kuzminska[73] also reported that nitrogen given along with phosphorus stimulated both the alkaloid production and capsule yield. Potassium slightly depressed morphine content. Nowacki et al.[70] also indicated that nitrogen application increased the amino acid, phenylalanine, which is a substrate for alkaloid, hence increased alkaloid content. Kuzminska[74] reported that other macronutrients like magnesium or calcium also increased the yield and morphine contents. Magnesium deficiency brings about a marked elongation of stem and early flowering without a decrease in morphine contents, while calcium deficiency causes a drop in alkaloid contents. Sodium favors flowers and capsule development and this increases output of morphine.[71]

Spasenovski[80] observed that application of salts like sodium chloride and sodium sulphate increased the morphine production.

Zogg[75] reported that heart rot disease of poppy plants is due to boron deficiency. To overcome this deficiency, Laughlin[76] observed that foliar spray of sodium borate increased capsule and seed yield. Ramanathan[77] also reported that foliar application of micronutrients or mixture of micronutrients (tracel) increased morphine contents and opium yield. Michna and Szwadiak[78] observed that boron combined with NPK gave the highest yields and morphine content.

Soil pH

The best for growth of poppy plant was considered to be pH_7. At higher or lower pH growth was badly affected.[79]

REFERENCES

1. Bare, C. E., Toole, V. K. and Gentner, W. A. Temperature and Light Effects on Germination of *Papaver bracteatum, P. orientale* and *P. somniferum*. Planta Med., 34(2), 1978, p. 135.

2. Volovich, E. M. and Griff, V. G. Minimal Temperature for Seed Germination. Fiziol Rast., 21(6), 1974, p. 1258.

3. Firsova, M. K. Effect of Low Temperature and Freezing on the Germinating Capacity of Seeds of Opium Poppy. (*Papaver somniferum*). Trudy Prikl. Bot. Ser., 4, Semenov., 2, 1937, p. 121.

4. Anderson, G. and Olsson, G. The Frost Hardiness of Spring Sown Oil Crops, Sveriges Utsadesforen. Tidskr., 60(2), 1950, p. 225.

5. Nesterenko, V. G. Germination of Seeds Stored Under Laboratory Conditions. (Rus) Bjull. Glavn. Bot. Sada., 36, 1960, p. 99.

6. Ghisleni, P. L. Contributions to the Knowledge of the Effects of Ultrasonics on Higher Plants. Ann. Accad. Agric. Torino., 98, 1955, p. 63.

7. Ohashi, H. The Geographical Variability of Chemical Substances in Plants in Relation to Vernalization, (Rus) Agrobiologija, 2, 1962, p. 268.

8. Ohashi, H., Tagawa, A., and Chiang, S. M. A Study of Vernalization of opium poppy, 2. Effect of Temperature of Treatment and Water Content of Seeds on the Development and Yields of Opium and Its Composition. Bot mag (Tokyo), 77, 1964, p. 300.

9. Hirosi, O. Results of Experiments on the Vernalization of Medicinal and Oil Plants, (Rus) Agrobiologija, 3, 1960, p. 427.

10. Lecat, P. Influence of Some Physiological Actions on the Biogenesis and Migration of Morphine in Opium Poppy Pods, Bull. Soc. Franc. Physiol. Veg., 7, 1961, p. 43.

11. Zsolt, J. The Effect of Eosin on Germinating Seeds (Fr) Borbasia, 7(1/10), 1947, p. 27.

12. Grover, I. S. and Dhanju, M. S. Effect of Gamma Radiation on the Germination of *Papaver somniferum* and *P. rhoeas*, Indian J. Plant Physiol., 22, 1979, p. 75.

13. Ghiorghita, G., Elvira, I., Gille, V. and Ecaterina, J. Toth. Gamma Irradiation, Ethylmethansulfonate and Diethylsulfate Treatments Induced Changes in Morphine Content and Other Biochemical Parameters in *Papaver somniferum*, Rev. Roum Biol. Ser. Biol. Veg., 27(2), 1982, p. 121.

14. Michalski, T. Effect of Ionizing Radiation on Morphine Content of Poppies (*P. somniferum* L.) (Pol. Summ. Ger.) Biul. Inst. Roslin. Leczniczych., 6(2), 1960, p. 169.

15. Fujita, S., Kawatani, T. and Kuriharo, K. Destructive Effect of Vapor Heating and Gamma Ray Radiation on Germination of *Papaver somniferum* and *Cannabis sativa.*, Bull. Nat'l. Inst. Hyg. Sci., 85, 1967, p. 68.

16. Husain, A. and Sharma, J. R. The Opium Poppy, Central Institute of Medicinal and Aromatic Plants, Lucknow (India), 1983, pp. 1-167.

17. Fairbairn, J. W. and Kapoor, L. D. The Laticiferous Vessels of *Papaver somniferum* L., Planta Medica., 8(1), 1960, p. 49.

18. Cheema, H. S., Chakravarti, B. P. and Lal Thakore, B. B. Effect of Opium Poppy Seed Treatments with Dithiocarbonates on Germination, Host Vigor and Viability of Seeds in Storage, Indian J. Plant Pathol., 5(2), 1987, p. 180.

19. Gorgiev, Milko and Dragana Andrievic. Effect of Chlorocholine Chloride on Yield and Morphine, Phosphorus and Potassium Content of Poppy, Agrohemija, 0(4), 1988, p. 291.

20. Smogrovicova, H., Kovacs, P. and Mikulas, P. Poppy Seed Germination:2 Shikimate Dehydrogenase (EC1.1.1.25) in Seedlings of P. Somniferum Cultivar Dubsky, Bioligia (Bratisl), 36(12), 1981, p. 1165.

21. Zemanek, J. and Ambrozova, J. The Study of the Effect of Amitrol on the Respiration and Activity of Some Enzymes in Poppy Plants (*P. somniferum* L.), (Sum Czech & Rus.) Biol. Pl., 9(4), 1967, p. 270.

22. Kohji, J., Hagmoto, H. and Masuda, Y. Georeaction and Elongation of the Flower Stalk in a Poppy, *Papaver rhoeas* L., Plant Cell Physiol., 17, 1979, p. 23.

23. Kohji, J., Nishitanj, K. and Masuda, Y. A Study on the Mechanism of Nodding Initiation of the Flower Stalk in a Poppy: *Papaver rhoeas*, Plant Cell. Physiol., 22(3), 1981, p. 413.

24. Sugawara, T. Effect of Chemical Growth Substances on the Bending of Flower Stalks in Papaver, Proc. Imp. Acad. Japan, 18, 1942, p. 89.

25. Benada, J. Redox Potential Gradient in the Flower of *Papaver somniferum.*, Biol. Plant Acad. Sci. Bohemostov., 9(3), 1967, p. 202.

26. Free, J. B. Insect Pollination of Crops. Academic Press, New York, 1970, p. 544.

27. Veselovskaya, M. A. "The Poppy." Amerind Publishing Co. (Russ. translation), New Delhi, 1976.

28. Coic, Y., Lesaint, C., Papin, J. L. and Lelandias, M. The Organic Acids and Mineral Substances in the Organs of Opium Poppy; Their Evaluation in the Reproductive Parts during Seed Development. Ann. Physiol. Veg., 10(1), 1968, p. 29.

29. Luthera, R. and Neelam, S. Changes in Fatty Acid Composition Accompanying the Deposition of Triacylglycerols in Developing Seeds of Opium Poppy (*P. somniferum* L.), Plant Sci. (SHANNON), 60(1), 1989, p. 55.

30. Marin, P., Sajdl, V., Kapor, S., and Tatic, B. Fatty Acid Composition of Seeds of the Papaveraceae and Fumariaceae, Phytochemistry (OXF), 28(1), 1989, p. 133.

31. Gentner, W. A., Taylorson, R. B. and Borthwick, H. A. Responses of Poppy, *Papaver somniferum* to Photoperiod. Bull. Narcotics, XXVII, No.2, 1975, p. 23.

32. Mika, E. S. Studies on the Growth and Development and Morphine Content of Opium Poppy, Bot. Gaz., 116, 1955, p. 323.

33. Khlebnikova, N. A. Growth and Development of White Poppy on Varying Daylength, Comp. Rend. (Dokl) Acad. Sci. Urss., 32(7), 1941, p. 503.

34. Lona, F. and Bocchi, A. Vegetative and Reproductive Development of Some Longday Plants in Relation to the Effect of Gibberellic Acid (Ital. Summ. Eng.), Nuovo Giorn. Bot. Ital., 63(4), 1956, p. 469.

35. Zebrak, E. A. The Effect of Polyploidy on Photoperiodic Reaction of Opium Poppy and Love-in-a-mist, (Rus.) Madl. Glav. Bot. Sada., 70, 1968, p. 105.

36. Felklova, M. and Levakova, K. The Influence of Light on Growth and Contained Substances in *P. somniferum*, (Ger. Sum. Rus. & Eng.), Acta. FAc. Pharm, Univ. Comeniana. 14, 1967, p. 7.

37. Ravenna, C. The Effect of Wounds on Morphine Production in Poppies (Ital.), Staz. Sperim. Agrar. Ital. 57, 1924, p. 5.

38. Prokofiev, A. A. and Kats, K. M. Transpiration of Fruits and Inflorescences Depending on the Intensity of Metrological Factors and the Age of Plants (Rus.), Fiziol. Rast., 10(2), 1963, p. 204.

39. Prokofiev, A. A. and Godneva, M. T. Significance of Photosynthetic Activity of Opium Poppy Fruits for Development of Seed and Fat Accumulation in Them, Proc. Acad. Sci. USSR. Bot. Sci. Sect., 114(1/6), 1957, p. 99.

40. Prokofiev, A. A. and Godneva, M. T. The Role of Photosynthetic Activity of the Fruits of the Oil Poppy in Seed Development and Oil Accumulation (Rus.) Dokl. Acad. Nauk. SSR., 114(2), 1957, pp. 438-441.

41. Rustembekov, S. S. Seasonal Changes in the Photosynthetic Capability of the Opium Poppy, (Rus.), Trudy Kirg. Univ. Ser. Biol. Nauk. 1967, pp. 23-25.

42. Zoschke, M. Mineral Nutrition and Morphine Formation in *Papaver somniferum* L. (Ger., Sum. Eng.), Z. Acker-Pflanzenbau., 116(4), 1963, pp. 317.

43. Spasenoski, M. and Jordanovsk, V. Effect of Increasing Doses of Sodium Chloride and Sodium Sulfate on the Yield of Dry Substance and Content of Some Elements in Opium Poppy (*P. somniferum* L.), Acta. Biol. Med. Exp., 13(1), 1988, p. 37.

44. Pfeifer, S. and Heydenreich, K. Die Akkumulation der Mohnaalkaloide Zwischen Blute and Biologischer Reife. Ein Beitrag Zum Problem Gewinnug von Alkaloiden aus Grummohn. Pharmazie., 17, 1962, p. 107.

45. Heydenreich, K. and Pfeifer, S. Metabolism of Alkaloids in *Papaver somniferum* L. IV The Distribution of Alkaloids in Plants Prevented to Set Seeds by Cutting of the Flower, Sci. Pharmaceut., 30(1), 1962, p. 17.

46. Bunting, E. S. Changes in the Capsule of *P. somniferum* between Flowering and Maturity, Ann. Appl. Biol., 51, 1963, p. 459.

47. Johri, M. M. and Maheshwari, S. C. Studies on Respiration in Developing Poppy Seeds. Pl. Cell. Physiol., 6(1), 1965, p. 61.

48. Maheshwari, S. C. and Johri, M. M. Respiration in the Developing Ovules of Poppy (*P. somniferum* L.), Naturwissenschaften, 50(23), 1963, p. 718.

49. Johri, M. M. and Maheshwari, S. C. Changes in the Carbohydrates, Proteins and Nucleic Acids during Seed Development in Opium Poppy., Pl. Cell. Physiol., 7(1), 1966, p. 35.

50. Balatkova, V. and Tupy, J. The Stimulatory Effect of Uracil and 5-bromouracil on the Seed Set of *P. somniferum* L., Biol. Pl. (Praha)., 14(2), 1972, p. 140.

51. Guillaume, A. and Faure, J. On the Variation in Morphine Content in Poppy Capsules during the Period of Maturity and during Storage (FR). Ann. Pharm. Franc., 1(1), 1946, p. 160.

52. Gracza, P. and Verzar, G. Effect of Growth Regulators on the Development and Alkaloid Content of Poppy Capsules, Acta Agron. Acad. Sci. Hung., 19(3/4), 1970, p. 406.

53. Bernath, J. and Vagujfalvi. Effect of CCC and DMSO on Poppy, Herb. Hung., 9(3), 1970, p. 49.

54. Zhdanova, L. P. and Rusova, M. I. Effect of Malic Hydrazide on the Accumulation of Storage Substance in Seeds, Fiziol. Rast., 26(2), 1979, p. 428.

55. Ramanathan, V. S. Effect of Plant Growth Regulators on Yield of Opium and Its Morphine Content in Opium Poppy, Indian J. Agric. Res., 12(4), 1978, p. 246.

56. Lecat, P. Effect of Spraying Maleic Hydrazide and Potassium Sulphate on Morphine Content of Capsules of Poppy (FR), Acad. Agric. France, Compt. Rend., 45(12), 1959, p. 592.

57. Gracza, P. Effect of Cytokinins on the Development of Poppy Capsule, (Hung. Summ. Eng.), Bot. Kozlem., 56(4), 1969, p. 263.

58. Lecat, P. Effect of Gibberellin on the Opium Poppy (FR), Bull. Bot. France, 107, 1960, p. 317.

59. Zsadon, B. Investigation of Agueous Extracts Obtained from Poppy Heads (Hung; Summ. Ger.), Magyar Kem. Folyoirat., 66(9), 1960, p. 347.

60. Heeger, E. F. Varietal Studies for Opium Production in Germany, (Ger.) Forschungsdienst., 8, 1939, p. 508.

61. Zdenda, H. The Breeding of Poppies for the Increased Content of Morphine (Czech. Summ. Russ. & Ger.), Preslia., 27(4), 1955, p. 368.

62. Felklova, M. and Mikulecka, J. Fluctuation of the Morphine Contents during the Vegetation of the Plant *Papaver somniferum* L. (Czech. Summ. Rus. Eng. & Ger.), Sborn. Ceskoslov. Akad. Zemed. Ved. Rostl. Vyroba., 4(2), 1958, p. 149.

63. Annett, H. E. Investigations on Indian Opium No. 2. The Effect of Environmental Factors on the Alkaloid Content and Yield of Latex from the Opium Poppy (*Papaver somniferum*) and the Bearing of the Work on the Functions of Alkaloids in Plant Life, Mem. Dept. Agric. India, Chem. Ser., 6(2), 1921, p. 61.

64. Annett, H. E., Sen, H. D. and Singh, H. D. Investigations on Indian Poppy No. 1. Non-environmental Factors Influencing the Alkaloidal Content and Yield of Latex from the Opium Poppy (*Papaver somniferum*), Mem. Dep. Agric. Chem. Ser., 4(1), 1921, pp. 1-60.

65. Heeger, E. F. and Schroder, H. Investigations of the Morphine Yield of *Papaver somniferum* L. under Central German Cultivation Conditions. Pharmazie, 14(4), 1959, p. 228.

66. Holloway, P. J. The Effects of Superficial Wax on Leaf Wettability, Ann. Appl. Biol., 63(1), 1969, p. 145.

67. Ramanathan, V. S., Sinha, A. K. and Vijaywargiya, S. K. Preservation of Morphine in Opium by Spraying Chemicals on Capsules and Latex of *Papaver somniferum*, Indian J. Agric. Res., 17(1/2), 1983, p. 79.

68. Felklova, M. and Vaverkova, S. The Yield of the Codeine from *Papaver somniferum* under Different Climatic Conditions, Farm. OBZ., 53(6), 1984, p. 251 (in Czech with Russ. Engl. and Germ. Summ.).

69. Sheberstov, V. V., Fomenko, K. P., Zhuravlev, Yu. P., Arsyukhina, N. N., Gindich, N. N., Poludennyi, L. V., Nerterov, N. N., Naumova, G. E. and Fonin, V. S. Fertilizers, Yields and the Active Principle Content of Medicinal Plants, Hort. Abstr., 44(5), 1972.

70. Nowacki, E., Jurzysta, M., Gorski, P., Nowacka, D. and Waller, G. R. The Effect of Nitrogen Nutrition on Alkaloid Metabolism in Plants, Biochemic and Physiologie der Pflanzen, 169(3), 1976, p. 231.

71. Costes, C., Milhet, Y., Gandilion, C. and Magnier, G. Mineral Nutrition and Morphine Production in *P. somniferum.*, Physiol. plant., 36(2), 1976, p. 201.

72. Kinoshite, K., Nakagawa, Y. and Isaka, H. Studies on the Effect of Nitrogenous Manures upon the Growth and Yield of Opium Poppy, Bull. Nat'l. Hug. Lab., 77, 1959, p. 267.

73. Kuzminska, K. Effect of N. P. and K fertilizing on the Contents of Some Principle Alkaloid in the Capsule of *P. somniferum* L., Herba Pol., 19(3), 1973, p. 256.

74. Kuzminska, K. Effect of Magnesium Fertilization at Two Calcium Levels and Soil Moisture on Crop and Morphine Content in Opium Poppy, Herba Pol., 18(3), 1973, p. 266.

75. Zogg, H. Opium Poppy Heart Rot and its Control, Flungel Landev Vers. Anst. Zurich Oerliken., 14, 1944, p. 4.

76. Laughlin, J. C. The Boron Nutrition of Poppies (*Papaver somniferum* L.) on Chernozem and Alluvial Soils of Tasmania, Planta med., 36(3), 1979, p. 245.

77. Ramanathan, V. S. Effect of Micronutrients on the Yield of Opium and Its Morphine Contents in Opium Poppy, Indian J. Agric. Res., 13, 1979, p. 85.

78. Michna, M. and Szwadiak, J. Effect of Boron and Nitrogen Fertilizing upon the Crop and Morphine Content in Poppy Heads of *Papaver somniferum* L. Variety "Niebreski K. M." (Pol. Summ. Rus. & Eng.), Biul. Inst. Roslin. Lecznichzych., 10(2-3), 1964, p. 138.

79. Kinoshita, K., Nakagawa, Y., Isaka, H. and Komine, T. Studies on the Effects of Soil Acids upon Growth and Yield of Opium Poppy (*Papaver somniferum* L.), Bull Nat'l. Inst. Hyg. Sci., 80, 1962, p. 158.

80. Spasenovski, M. Effect of Sodium Chloride and Sodium Sulphate on Alkaloid Production and Mineral Content of Poppy, (*P. somniferum*), Hort. Abstr., 50(7), 1980, p. 4841.

81. Malinina, V. M. and Ivanova, R. M. Morphine Accumulation in Some Poppy Cultivars in Different Climatic Zones, Hort. Abstr., 44(12), 1974, p. 889.

82. Bernath, J. and Tetenyi, P. The Effect of Environmental Factors on Growth Development and Alkaloid Production of Poppy (*P. somniferum* L.) Response of Day Length and Light Intensity, Biochem. Physiol. der Pflanzen, 174(5-6), 1979, p. 468.

83. Farkas-Riedel, L. Studies on the Changes in Peroxidase Activity in *Papaver somniferum* L. Varieties During Ontogenesis, Acta. Agron. Acad. Sci. Hung., 18, 1969, p. 317.

Chapter 6

Anatomical Studies

Anatomical studies in general on plants of Papaveraceae have been described by Metcalf and Chalk[1] but no details are covered by them for *Papaver somniferum* L. in particular. The stem in general, in transverse section, exhibits a single ring of widely spaced bundles which are mostly collateral. Sometimes several rings of bundles are present in *Papaver*. Scanty uniseriate, biseriate, or multi-seriate hairs and occasionally shaggy hairs are present. In transverse section the petiole commonly exhibits an arc of vascular bundles not accompanied by sclerenchyma. Ranunculaceous stomata are present. The latex of various colors is present in articulated laticiferous tubes or laticiferous cells or sacs. Alkaloids of morphine or codeine, etc., are present in the latex.

LEAF

Usually dorsiventral. Shaggy hairs are present in *Papaver*. The epidermis is frequently covered with wax. Stomata is present on both surfaces in *Papaver pilosum* Sibth ex. Smith, and *P. spicatum* Bois., but confined to lower surfaces in other genera and species. Hydathodes occur in groups on the lower surface of the teeth at the margin of the leaf in certain species of *Papaver*. Mesophyll generally includes one to several layers of palisade cells but is not distinctly differentiated into palisade and spongy regions in *Hesperomecon platyslemon* Green, *Meconella Californica* Torr., *M. oregana* Nutt, and *Papaver somniferum* L. Petiole in transverse sections through the distal end exhibits an arc of vascular bundles in *P. dubium* and few other genera. Bundles are sometimes very close together. Sclerenchyma is generally scanty or absent from the petiole.

Latex is generally present throughout the plant either in (1) articulated laticiferous tubes or (2) laticiferous sacs consisting of elongated cells either solitary or arranged in longitudinal rows. Articulated laticiferous tubes are recorded in *Papaver* and a few other genera. Color and consistency of latex is variable, being white in *Papaver,* lemon yellow in *Argemone* and orange in *Chelidonium,* etc.

STEM

A collenchymatous exodermis and peripheral part of the cortex is followed by the endodermis. Pericycle is weakly sclerosed in *Papaver somniferum* L. and *P. orientale* L. Most members of the Papaveraceae are provided with a single ring of collateral vascular bundles. Numerous bundles are arranged in concentric zones in *Papaver somniferum* L. and *P. orientale* L.

PEDUNCLE

The anatomical structure of the peduncle presents usually with four to five vascular bundles in *Papaver dubium* L. and *P. somniferum* L. Cortex is narrow, component cells generally with cellulose walls, becoming hollow. Vascular bundles are widely separated and arranged in a single circle, their number varying according to species, as well as in different individuals of the same species, or at different levels in a single peduncle. Xylem groups are not generally u-shaped.

One of the most interesting anatomical features is the tendency for the vascular bundles in a few species to be scattered. The xylem groups are also occasionally u-shaped. Both of these characters suggest affinities with the Ranunculaceae and therefore with some of the monocotyledons. The laticiferous elements are generally believed to be homologous with the secretory cells of the closely related Fumariaceae.

The ontogeny, structure, distribution, and functions of laticiferous tubes or laticiferous sacs have not been described.[1] Because they play an important role in the life history of *Papaver somniferum* L., a detailed information on this specialized tissue is discussed herein.

THE LATICIFEROUS VESSELS

Laticifer is the generic term used to cover all types of specialized cells and vessels which produce latex–a substance consisting of liquid matrix regarded as cell sap, with organic particles in suspension.[2] Substances like carbohydrates, organic acid, salts, alkaloids, sterols, fats, tannins, and mucilages are found within the laticifers. Resins and rubber are characteristic components of latex in many plants but in others the latex may consist of terpenes and the rubber may be lacking. The walls of laticiferous vessels in general are primary, soft, and apparently plastic. They are not thicker than the parenchymatous cell wall but the thickness may increase slightly with increasing age. Some band- and knot-shaped thickenings in the walls of laticifers are reported at the base of the stem in *Argemone* by DeBary.[3] The thickenings of the walls may be uneven but no primary pit-fields have been seen with certainty.

Within the wall of the laticifers neither protoplasm nor nuclei are seen. However, many forms of coagulated finely granular latex resemble coagulated protoplasm or their remains but no anatomical evidence could be seen. The nature of the protoplast of laticifers has been imperfectly investigated and according to Esau,[2] its study is as difficult as that of the sieve element protoplast.

The common concept is that laticifers maintain a living protoplast; that the nuclei remain in this protoplast upon the maturation of the elements and that the cytoplasm occurs as a parital layer enclosing a vacuole composed of latex. Frey-Wyssling[4] reported that in the young laticifers the nuclei can be distinguished but later the dense latex obscures their visibility. Esau[2] pointed out the difficulty in obtaining a proof of the presence of parital cytoplasm. As pointed out by Bonner and Galston,[5] there is no clear demarcation between cytoplasm and the vacuole in mature laticifers and in the sectioned material there is considerable displacement of contents.

Laticifers have been found only in Angiosperms, and that, too, only in relatively few scattered families. On the basis of structure and ontogeny Easau[2] has classified laticifers as shown in Table 6.1.

The articulated laticifers are compound in origin and consist of longitudinal chains of cells in which the end walls separating the cells become perforated or are completely absorbed. This type of

TABLE 6.1. Laticifers

Articulated Laticiferous Vessels		Non-articulated Laticiferous Cells	
Articulated anastomosing	*Articulated non-anastomosing*	*Non-articulated branched*	*Non-articulated unbranched*
viz. *Chichorium;*	viz. *Ipomoea;*	viz. *Euphorbia;*	viz. *Vinca;*
Lactuca;	*Convolulus;*	*Asclepia;*	*Urtica;*
Taraxacum;	*Chelidonium;*	*Cryptostegia;*	*Cannabis.*
Tragopogon;	*Achras;*	*Nerium;*	
Lobelia;	*Allim;*	*Ficus.*	
Carica;	*Musa.*		
Papaver;			
Argemone;			
Hevea;			
Manihot.			

Easau, K. Plant Anatomy. John Wiley and Sons, Inc., New York, 1953.

laticifer is often called laticiferous vessels. The non-articulated laticifers originate from single cells, which, through continued growth, develop into tube-like structures, but typically they undergo no fusion with other similar cells. Being simple in origin, they are often called laticiferous cells. Due to variations in the two types of laticifers, they are subdivided into two groups each. Some articulated laticifers consist of long chains or compound tubes connected with each other laterally and forming anastomoses and others not connected with each other laterally. The types are called articulated anastomosing and articulated non-anastomosing laticifers. The non-articulated laticifers also exhibit some variations in their structure. Some develop into tubes which branch off repeatedly and others develop into long and more-or-less straight tubes without any branching. These are designated as non-articulated branched and non-articulated unbranched laticifers.

From the examples of various types of laticifers found in different families it is obvious that the type of laticiferous element is not constant in a given family. Euphorbiaceae exhibits articulated and non-articulated laticifers distributed in different genera. Papaveracea exhibits the articulated anastomosing and articulated non-anastomosing laticifers distributed in different genera of the plants, viz. *Papaver* and *Argemone* for the former, and *Chelidonium* for the latter, but two types of laticifers have not been reported in the same plant.

A review of the literature of the laticiferous vessels of Papaveraceae shows that very little attention has been devoted to opium poppy. Trecul[6] first reported the presence of laticifers which he called lacteal ducts in *Papaver somniferum* occurring under the phloem tissue. Leger[7] made a more systematic investigation on the topography and histology of lacteal ducts in some plants of Papaveraceae but he limited his investigations to the stem and leaf. He observed that in many species the lacteal ducts are situated in a convex zone at the limits of the primary and secondary phloem of the stem and leaf bundles. He reported some more external lacteal ducts in the median region of the primary phloem in larger vascular bundles of *Papaver somniferum*. He described the ducts as wide with rounded orifices but no direct contact communication of lacteal ducts with xylem vessels existed either in the stem or in leaf as reported earlier by Trecul.[6] He did not investigate lacteal ducts in the flower or capsule but reported their course in the stem as just the same as that of vascular bundles in which they are contained.

Regarding the development of laticifers, Esau[2] confirms the earlier work of DeBary[3] on Papaveraceae and concludes that single files of cells are transformed into tubes by perforation of the end walls viz. Chelidonium; or by a partial or complete resorption of end walls and a development of lateral anastomoses joining the tubes with each other, viz. *Papaver, Argemone*.

The presence of laticifers has been scarcely illustrated in *Papaver somniferum* except by Leger[7] in the stem and by Tschirch and Oesterle[8] and Fedde[9] in the capsule. The work of Bersillon[10] deals more with the developmental anatomy of apices in the Papaveraceae and does not discuss the laticifers in particular.

Trecul[6] observed long narrow cells in the young roots which were laticiferous cells in the process of formation but not yet joined up into a laticiferous system.

From the review of the earlier literature it appears that little work has been done on the development of laticiferous vessels in the floral parts of the opium poppy. Similarly, information on the development of laticifers in the young seedling is lacking. As the appearance of alkaloids is intimately associated with the presence of laticifers, it would be interesting to find out the first appearance of these vessels in the growing seedling.

Anatomical evidence for the presence of laticifers, specially in some organs of the plants, viz. growing seedlings or floral parts (especially in stamens), has altogether been lacking. Phytochemical work for the presence or absence of alkaloids in these organs has been done mostly on a qualitative basis.

GENERAL DESCRIPTION OF THE CULTIVATED PLANT

Since *Papaver somniferum* L. is a cultivated plant, it shows considerable variations in the color of flower, seeds, and the shape of the capsule. However, the general appearance of the plant does not vary so much (Figure 6.1).

The plants of *Papaver somniferum* L. grown for experimental purposes may be described as annual herbs with thick tapering root and stems reaching a height of 3-4 ft. erect, cylindrical, solid, and quite smooth. The leaves are large, numerous, alternate, sessile, and clasp the stem by a cordate base. The buds are ovate and drooping but the flowers are erect, solitary, and large. Two sepals, which are green, broad, and quite smooth, disarticulate and are pushed away as the flower opens. The four large petals are decussate, the outer two are wider and much overlapping the slightly narrower inner ones. They are concave, undulated, with numerous closely placed veins radiating from the stiff, thick, wedge-shaped base. They are pure snow white and glossy.

The stamens are numerous, hypogynous, and inserted in two to three rows on the undersurface of the dilated thalamus. Filaments are long, flat, and ribben shaped, slightly dilated at the top. Anthers are linear and attached by a narrow base to the filaments. They are cream colored, wavy, and are twisted after dehiscence.

The ovary is large and globular but contracted below into a neck (gynophore) which again dilates to form the receptacle, and this also narrows off below into the pedicel. The latter is quite smooth

FIGURE 6.1. *Papaver somniferum* L. (opium poppy), showing the whole plant, floral parts, capsule, seed, and transverse section of ovary. Reprinted from Kirtikar and Basu's "Indian Medicinal Plants" with permission from Bishen Singh and Mahmderpal Singh of Dehradum (India).

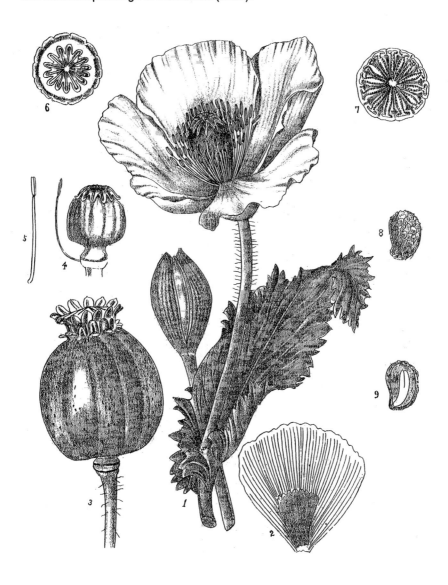

PAPAVER SOMNIFERUM, *LINN.*

and green in color. The ovary is one-celled and contains a large spongy parietal placenta which bulges out nearly to the center. The placentae are almost always equal to the number of stigmatic rays and bear numerous ovules over all parts of their surface. The stigma is sessile, peltate, and spreading over the top of the ovary with eight to 13 short, obtuse, oblong rays.

The fruit, a capsule, is usually more or less globular, supported on a neck (as in ovary), and crowned by persistent stigma. The pericarp is hard, smooth, dry, and brittle and brownish-yellow when ripe. It has one central cavity with dry papery placental plates reaching about half-way to the center.

Capsules of some varieties are dehiscent and of others indehiscent. The poppy grown above in the experimental plot was indehiscent. The seeds are small but numerous, reniform, and yellowish-white in color. The testa has a reticulated network and the embryo is slightly curved within the oily endosperm.

When the bud unfolds, the whole flower is open during the day, the sepals tend to drop first and the petals fall off after 24 to 48 hours. The stamens may persist for a short period after the petals drop but they soon dry up. The ovary, after fertilization and after the petals have dropped, develops in size and within two weeks assumes considerable dimensions when it can be considered fit for lancing. For the anatomical work described below, the pedicel and capsule (at week 2) were cut into different segments (Figure 6.2) and serial sections were taken.[11] Sampling of capsules from the experimental plot was done soon after the fall of petals and this date was called week 0 for this stage of development of the capsule and at weekly intervals afterward collection of capsule was made:

	Number of Days After the Petals Drop
Week 0	0
Week 1	7
Week 2	14
Week 3	21
Week 4	28

FIGURE 6.2. Diagram of pedicel and capsule at week 2, showing regions A to D, used for anatomical studies. (L. D. Kapoor)

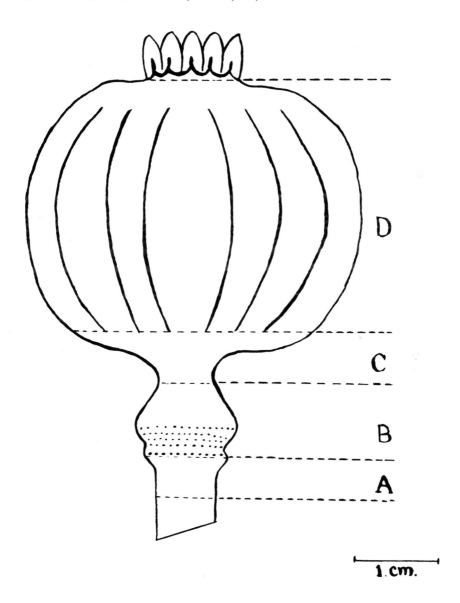

(a) *Course of vascular bundles and laticifers in the pedicel*

As the laticifers are closely associated with phloem tissue, their course can be easily followed by tracing the course of vascular bundles in opium poppy.

A transverse section of the pedicel at the distil end shows numerous vascular bundles arranged in concentric rings giving an appearance somewhat resembling a monocotyledonous stem (Figure 6.3(1)). The vascular bundles however are not scattered but arranged in two rings, an outer and an inner. The vascular bundles of the inner or central ring number 13 and are larger than the peripheral bundles of the outer ring which numbered 39-41. Approximately three peripheral bundles alternate with one central bundle. Of the three peripheral bundles the middle one is larger than the other two. The number of the central bundles is not constant for the species and it varies within the same population of the same species. In some of the plants it is nine and in others it is 11. The vascular bundles in the peripheral ring vary accordingly so that the feature of three peripheral bundles alternating with one central bundle seems to be pretty constant.

A thick cuticle on the epidermis and what appears to be hypodermis is seen in transverse and longitudinal sections (Figures 6.4 and 6.5). This tissue was called exodermis by Harvey-Gibson and Bradley,[12] who observed it in the stem of *Papaver somniferum* but could not confirm its presence in *P. rhoeas*. It is followed by five to seven layers of chlorenchyma consisting of polygonal or rounded parenchymatous cells with green chloroplasts and with prominent intercellular spaces. Immediately below this there are five to six layers of thick lignified pitted parenchyma. Harvey-Gibson and Bradley,[12] who called this the pericycle, as it forms a continuous ring outside the vascular bundles, stated that it was less sclerotic than that of *Papaver rhoeas*. The intensity of sclerosis however increased with the maturity of the plant. This tissue fuses with the cap of the vascular bundles of both the rings except for a few bundles of the central ring.

It is sometimes difficult to differentiate between the lignified parenchyma and the lignified phloem fibers in transverse section when they are fused together. Phloem fibers are however narrow and much longer in longitudinal section than the lignified parenchyma. Pits are

FIGURE 6.3. Transverse section through pedicel and lower thalamus. (Figure 6.2, region A) cent. b.–central vascular bundle; per. b–peripheral vascular bundles. Sections in ascending order 1-4. (L. D. Kapoor)

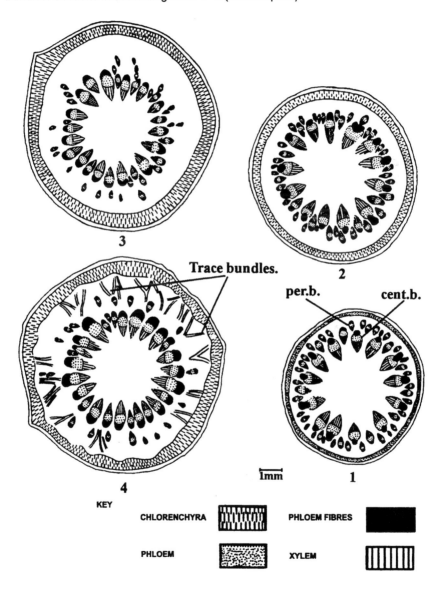

FIGURE 6.4. Transverse section of cortex and peripheral vascular bundle of pedicel. (Figure 6.3(1)). CAM–Cambium; CHL–chlorenchyma; CUT–cuticle; EP–epidermis; HYP–hypodermis; LAT–laticiferous vessels; PAR–parenchyma; PH–phloem; PH.FB.–phloem fibers; XY.V.–xylem vessels. (L. D. Kapoor)

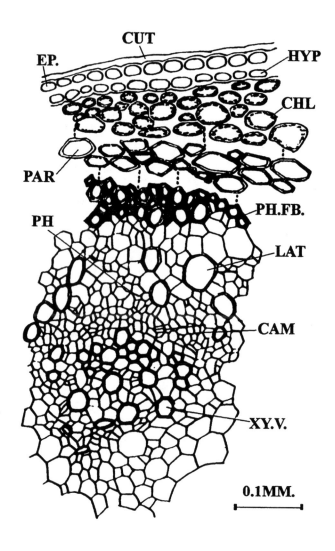

FIGURE 6.5. R. L. S. (radial longitudinal section) through pedicel. A—showing central and peripheral vascular bundles and cortex; B—section through phloem region of central vascular bundle; C—section through phloem region of peripheral vascular bundle. chl.-chlorenchyma; ep.—epidermis; lat.—laticiferous vessels; par.—parenchyma; ph.fb.—phloem fibers; s.p.—sieve plate; xy.v.—xylem vessels. (L. D. Kapoor)

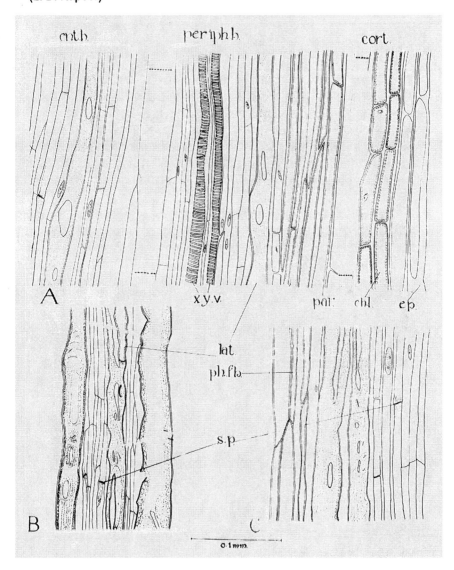

present on the fiber walls but they are not so prominent as on the walls of the lignified parenchyma cells. Immediately within the phloem fibers, the phloem parenchyma, sieve tubes, and companion cells are located. The laticiferous vessels are found within the phloem tissue. In transverse section the laticiferous vessels appear as an arc both in the central and peripheral bundles. But in the peripheral bundles they show the appearance of a convex arc (Figure 6.4) and in the central bundles a concave arc (Figure 6.6). The laticiferous vessels are articulated and show anastomosing within the bundle but laticifers of one bundle are never seen anastomosing with that of another. The position of laticifers in the peripheral and central bundles is seen sometimes near the limits of the primary and secondary phloem. In some of the central bundles the arc of laticiferous vessels extends in such a way as to surround the phloem fibers (Figure 6.6). The presence of granular latex can be detected easily in longitudinal sections when the vessels are stained with iodine or calco oil blue (Figure 6.5).

Phloem is followed by the cambial layer and xylem elements. Xylem consists of tracheids with annular or spiral thickenings and xylem parenchyma. Cells of medullary rays which occur in the interfasicular region have thick and pitted walls. The pith occupies the central place with thin-walled parenchymatous cells.

The size of vascular bundles varies a good deal but the diameter of the laticifers and sieve tubes exhibit marked variation at this stage of development. Table 6.2 shows the diameters of sieve tubes and laticiferous vessels in the central and peripheral bundles.

As the diameter of the laticifers is significantly higher than that of sieve tubes it is possible to use these measurements to distinguish sieve tubes from laticifers if it is not possible to do so by simple observation.

(b) *Course of vascular bundles and laticifers in the thalamus.*

The course of the vascular bundles as seen in the pedicel (Figure 6.3(1)) remains unchanged until it reaches the region of the thalamus (Figure 6.2(B)). The bundles in this region present a rather complicated course. They give off traces (Figure 6.7) which may later fuse with another trace from another bundle and the fused vascular strand may then branch again. Quite a number of these traces run vertically for a very short distance but soon bend toward

FIGURE 6.6. Transverse section of central vascular bundle (Figure 6.3(1)). CAM–cambium; LAT–laticiferous vessels; PAR–parenchyma; PH–phloem; PH.FB.–phloem fibers; XY.V.–xylem vessels. (L. D. Kapoor)

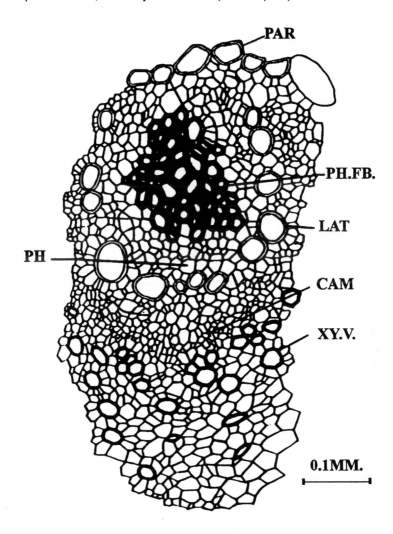

TABLE 6.2. Diameter in microns of laticiferous vessels and sieve tubes in vascular bundles of the pedicel

Serial section in ascending order	Central Bundles		Peripheral Bundles	
	Laticifers	Sieve Tubes	Laticifers	Sieve Tubes
1.	28.0	13.0	25.0	10.5
2.	30.1	12.2	25.7	10.1
3.	25.2	12.1	22.7	9.8
4.	29.5	13.3	23.5	9.3

the cortex and run into the floral leaves. These traces to sepals and petals consist of xylem, phloem, and laticifers, but not the phloem fibers. The bundles or traces always fuse before giving off new branches. Hence, in transverse section the number of bundles seen may increase or decrease as the pedicel is ascended. In ascending serial sections the development appears as follows: The two rings of bundles move toward the pith and the middle bundles of the peripheral ring move into the central ring which now contains about 26 vascular bundles (Figure 6.3(4)). Meanwhile, the remaining 26 small bundles originally in the peripheral ring move out toward the cortex and presumably pass into the sepals and petals. The 26 central bundles come together (Figure 6.8(5)), fuse in pairs (Figure 6.8(6,7)), and give off traces to the stamens. Later they divide once more to form a ring containing about 50 equal-sized bundles (Figure 6.8(8)) near the periphery. These then progressively fuse into groups of three to four bundles and form the main bundles of the capsule. The chlorenchyma tissue ceases in the region where most of the stamen traces are given off (Figure 6.8(6,7)), and it reappears when all the stamen traces have been given off (Figure 6.8(8)).

(c) *The course of vascular bundles and laticiferous vessels in the capsule.*

By the time the base of the capsule is reached, the bundles have fused to form 11 amphicribral bundles (Figure 6.9). This number varies with the plant, but the number entering the capsule is general-

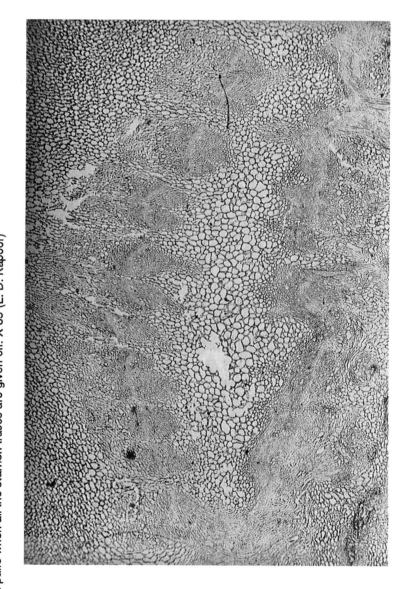

FIGURE 6.7. Photomicrograph of R. L. S. through thalamus showing the complicated course of trace bundles and their arrangement in pairs when all the stamen traces are given off. X 85 (L. D. Kapoor)

FIGURE 6.8. Transverse section through thalamus. (Figure 6.2, region B) tr.b.–trace bundles. Sections in ascending order 5-9. X 85 (L. D. Kapoor)

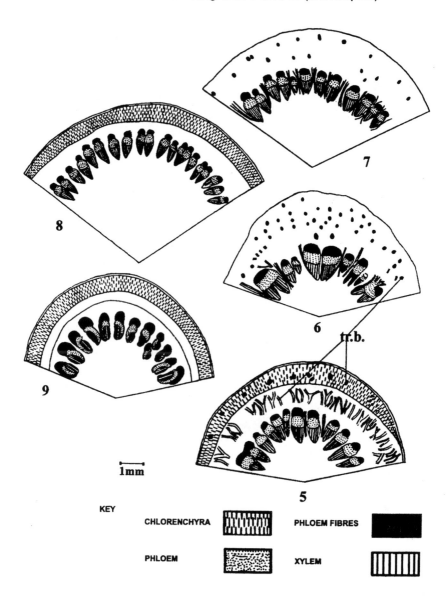

ly equal to the number of placental plates and stigmatic rays. They are the main bundles of the capsule and form the vascular framework. Their position is indicated by longitudinal depression on the outside of the capsule. Each depression with its bundle marks the position of a placental plate. These are called placenta bundles,[10] and the faintly convex regions between these bundles are known as valves. As these placenta bundles ascend from the narrow constricted region at the base of the capsule they enter the main capsule. A brief description of the structure of the capsule is given hereunder before describing the course of bundles.

Anatomically, the poppy head presents the typical structure of a fruit, the capsule, with well-defined layers of epicarp, mesocarp, and endocarp (Figure 6.10). The epicarp consists of epidermis with a thick layer of cuticle on the outside and collenchyma on the inside. Epidermis is a single layer of thick-walled parenchymatous cells with ranunculeous type of stomata. There is no hypodermis. Collenchyma consists of three to four layers of rounded or polygonal parenchymatous cells which are densely packed. This is followed by mesocarp. The mesocarp consists of large rounded parenchymatous cells containing chloroplast with intercellular spaces which increase in size from the periphery toward the interior (Figures 6.10, 6.11). The endocarp appears in transverse section as a single layer of tangentially elongated, lignified, thick-walled, and pitted cells lining the entire placenta from inside. Stomata can also be observed in this layer. Placental plates radiate as triangular lobes from the inner capsule wall carve out toward the hollow center, and these also consist of loose parenchymatous cells with large intercellular spaces. Their cells contain abundant starch grains. Numerous ovules arise on both sides of the placenta lobes (Figure 6.11).

The course of the vascular bundles and laticifers from the base of the capsule to its top can be followed by examining serial transverse sections. As the placenta bundles ascend, they bend gently toward the capsule wall and enter the base of the placental plate which is being carved out and projects into the hollow center as a triangular bulge (Figure 6.12(11)).As the placental bundles ascend, they give off branches toward the valves and to the placental plates. Vascular traces given off toward the valves may be called valve traces. These

FIGURE 6.9. Photomicrograph of T. S. (transverse section) through placenta showing the placenta bundle formed by the fusion of 5-6 vascular bundles. lat.–laticiferous vessels; ph.–phloem; xy.–xylem. X 85. (L. D. Kapoor)

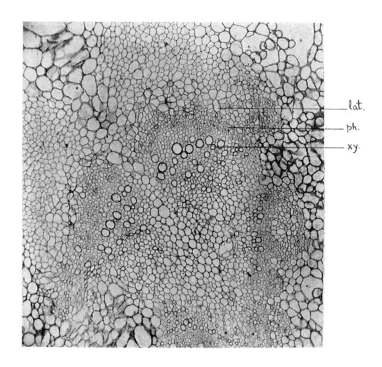

traces in the valves form the main source of latex in opium poppy (Figures 6.11, 6.13, 6.14(A) and (B)).

In the transverse section these traces are seen branching and anastomosing in the complicated way, but they are restricted to the mesocarp only and have not been seen in the epicarp (Figures 6.10, 6.11). As these traces are branched off from the placenta bundle they run somewhat parallel to the endocarp for a short distance, and then frequently the individual trace bends round on itself to produce a sickle-shaped appearance in transverse section so that a large portion of the trace runs somewhat parallel with the epicarp. The traces branch off freely and present a network traversing in all directions in the curved pericarp. These traces consist of xylem and phloem and prominent laticiferous vessels, but unlike the placenta bundles they

FIGURE 6.10. Photomicrograph of a transverse section through capsule wall. ep.–epicarp; meso–mesocarp; val. tr.–valve traces X 170. (L. D. Kapoor)

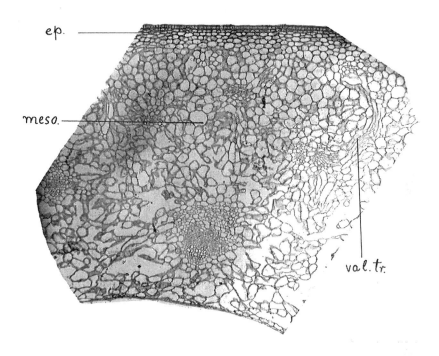

contain no phloem fibers. The orientation of xylem and phloem changes as the traces change their course. The xylem faces the endocarp when the traces are first given off from the placental bundle but later on when the traces bend and run parallel to the epicarp this orientation is inverted (Figure 6.13). The xylem consists of pitted vessels but the sieve tubes and companion cells are not well-developed in the phloem, however, the laticiferous vessels are well developed and are more prominent than phloem elements (Figure 6.15). Laticifers in this region show well-marked articulations and anastomosing with circular or oval orifices and granular contents and they are distributed throughout the mesocarp forming a network (Figure 6.14(B)). The density of articulations and anastomosing is perhaps the maximum here (Figure 6.14(A) and (B)). They show neither any association nor intercommunication with xylem vessels.

FIGURE 6.11. Photomicrograph of transverse section through capsule wall and placenta. pl.b.–placenta bundles; v.tr.–valve traces; pl.tr.–placenta traces. X 81. (L. D. Kapoor)

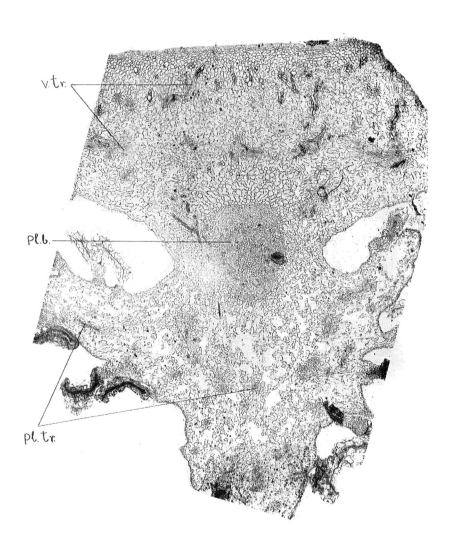

FIGURE 6.12. Transverse section through the base of capsule. (Figure 6.2, region C) tr.b.–trace bundles. Sections in ascending order 10-13. X 85. (L. D. Kapoor)

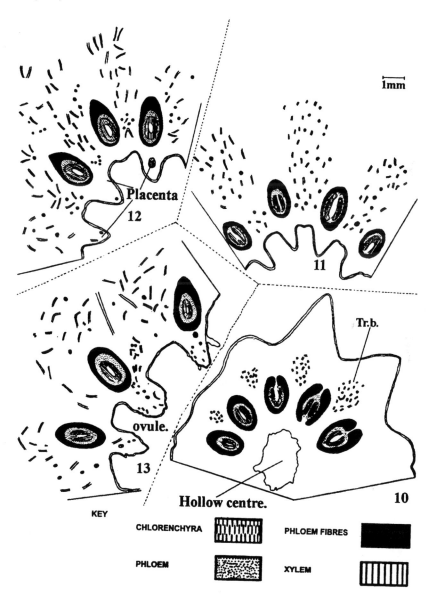

FIGURE 6.13. Photomicrograph of a transverse section through capsule wall showing a magnified valve trace with laticiferous vessels. L–laticiferous vessels; meso.–mesocarp; ph.–phloem; xy.–xylem vessels. X 220. (L. D. Kapoor)

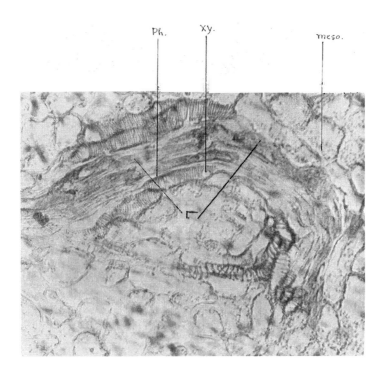

VASCULAR TRACES IN THE PLACENTA

The vascular traces which are given off by the placental bundle towards the placenta further give sub-branches which ramify in the entire tissue and these may be described as placental traces (Figure 6.14(A)). These traces are much smaller in diameter as compared with those in the valves and consist of one or two files of xylem tracheids with annular secondary thickenings and phloem elements with laticifers but with no phloem fibers. Some of these traces pass through the wall of the placenta into the base of the ovules (Figure 6.16(B) and (D)). These traces consist of a single file of xylem and phloem elements but no laticiferous vessels could be observed. It was observed that laticiferous vessels of placental traces terminate

FIGURE 6.14. A–R. L. S. of capsule in segment D (Figure 6.2) showing the placenta bundle giving traces to valves and placenta. B–Laticiferous vessels showing the anastomoses and orifices in a valve trace. C–Laticifers in a macerate of placenta trace near the margin of placenta where ovules are attached, showing terminal endings. D–Laticifers of a sepal trace in thalamus showing the protuberances in the process of anastomoses. B, C, D X 280. lat.–laticifers; mes.–mesocarp cells; pl.b.–placenta bundle; pl.tr.–placenta traces; v.tr.–valve traces; xy.v.–xylem vessels. (L. D. Kapoor)

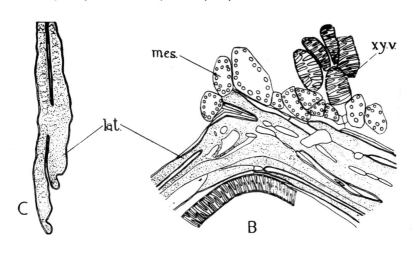

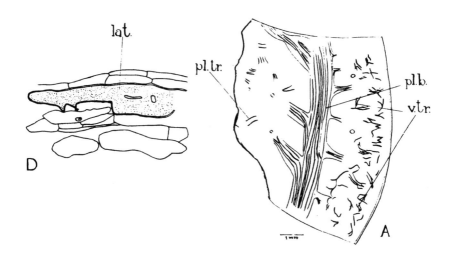

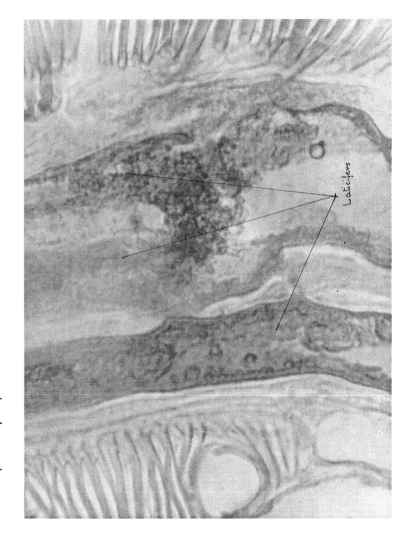

FIGURE 6.15. Photomicrograph of laticiferous vessels in valve traces showing anastomoses X 1700. (L. D. Kapoor)

Laticifers

FIGURE 6.16. Laticiferous vessels in placenta. A–transverse section of placenta trace; B–junction of ovule and placenta showing xylem element entering the ovule; C–termination of laticifers before the bundles enter the ovule; D–laticifers terminate while xylem elements continue toward the ovule. lat.–laticifers; ov.–ovule; par.–parenchyma cells of placenta; ph.–phloem; pla.–placenta; xy.–xylem. X 85. (L. D. Kapoor)

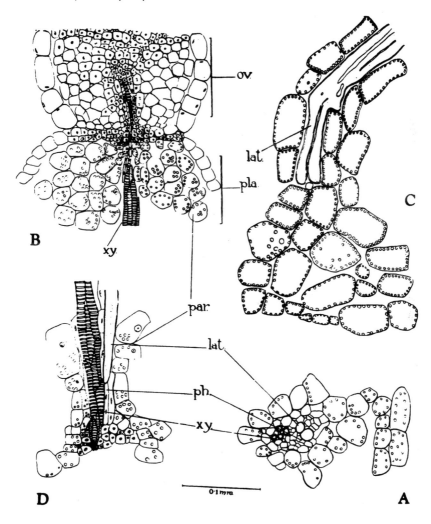

FIGURE 6.17. Transverse section through the capsule at different ascending levels. (Figure 6.2, region D) Sections in ascending order 14-18. X 85. (L. D. Kapoor)

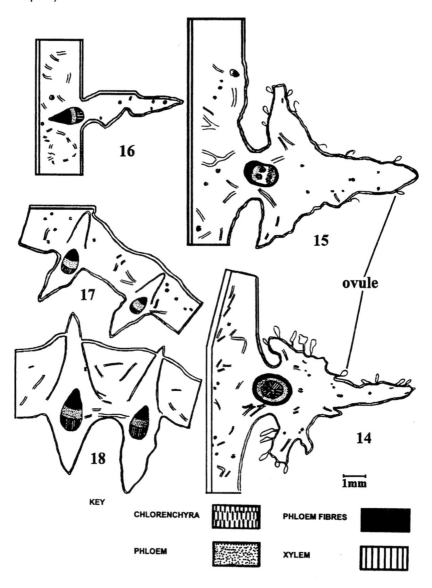

with rounded endings in placenta itself before the traces reach the wall of the placenta (Figures 6.14(C), 6.16(D)). The xylem and phloem of the traces, however, continue forward to pass through the wall and enter the base of the ovule (Figure 6.16(D)). The terminal endings of laticifers can be observed in a macerate from the marginal portion of placenta where the ovules are attached (Figure 6.14 (C)).

The course of the placental bundles and its branching to valves and placenta does not alter much throughout the length of the capsule. However, near the top of the capsule when the placental bundles move toward the stigma they become progressively reduced from an amphicribral to collatoral type of vascular bundles as they occur in the pedicel (Figure 6.17(14-18)).

STIGMATIC RAYS

The segments between the placental plates or valves are progressively reduced in size and finally disappear as the placenta bundles reach the distil end of the capsule. The stigmatic rays are developed from the placental plates and they are not connected in any way with the valves.

Each placenta bundle bifurcates when it reaches the top of the placental plate and each bifurcation fuses with another of the next placental bundle and the fused bundle enters the stigmatic ray where it gives off numerous branches. The vascular traces in the stigmatic rays carry the laticiferous vessels associated with the phloem.

It would thus be seen that the placenta bundles form the main vascular frame of the capsule. The valve traces which run in the capsule wall and anastomose freely arise as branches from the placenta bundles. The vascular supply to the placenta and ovules also arises from the placenta bundles and so do the vascular traces in the stigmatic rays. The phloem fibers which form a characteristic cap in the vascular bundles of the pedicel and later in the placenta bundles are not seen either in valve or placenta traces or stigmatic rays but the laticiferous vessels are present in all the traces. The laticiferous vessels end in the placenta and do not accompany the vascular traces into the ovule.

COMPARATIVE ANATOMY OF THE CAPSULES

(a) *Development in size*

The capsules attain a considerable size soon after the drop of petals. The seeds sown in April germinate to flowering stage after about 76 days and the weekly collections are made after the fall of petals. For determining the size, five capsules representing each stage are cut in two medium vertical halves and the surface area of each half worked out. The average area for each stage is shown in Table 6.3.

These stages of development are illustrated in Figure 6.18. It is obvious from the above that a rapid development of size takes place after the drop of petals in the first week. During the period from Week 1 to Week 2 the rate of development has slowed down but still there is an increase in size. During the later stages of development there is no increase in size. The maximum growth period, therefore, seems to be reached two weeks after petals fall.

This development is seen in all parts of the capsule, viz. in the pericarp, placenta, etc., but the mesocarp shows more development than the other layers of the pericarp.

(b) *Development of the vascular bundles*

There is no change in the general arrangement and constitution of the vascular bundles during the course of development of capsule from Week 0 to Week 3 or after. Lignification is, however, observed in the phloem fibers of the placenta bundle and the endocarp of the capsule which could not be seen in Week 0. The intensity of lignification increases as the capsule develops from Week 1 to Week 3. In the pedicel also the lignification of phloem fibers and of the parenchymatous tissue under the chlorenchyma is seen in Week 1 and its intensity increases as the age of the capsule develops. Cuticle layer also exhibits increasing thickness as the capsule matures. The placenta bundle shows a little increase in its diameter as the capsule matures from Week 0 to Week 2, but in the later stages no increase was noticeable (Table 6.4). The valve traces, however, show increased branching and anastomosing in the mesocarp as the capsule

TABLE 6.3. Average area of medium vertical section of capsules at different stages of development

Stage of development	Number of days after germination	Average area in Sq. cms.
Week 0	78	3.6-4.2-4.4
Week 1	85	14.2-14.6-14.9
Week 2	92	17.4-17.6-17.8
Week 3	99	17.0-17.2-17.4
Week 4	106	17.0-17.2-17.4

matures. These traces, which were lying deep in the mesocarp in Week 0 as seen in transverse sections, draw nearer to the epicarp as the fruit develops. This approach to the periphery is possible due to the increased branching of the valve traces which results in reaching their maximum density in Week 2. Branching of valve traces occurs in all directions in the mesocarp, but as already stated, in no case were these traces seen in the epicarp.

The valve traces show development of some xylem tracheids into xylem vessels with secondary thickenings as the capsule matures.

(c) *Development of laticiferous vessels in the ovary*

Anatomically, the basic constitution and the course of vascular bundles in ovary is not different from the other parts of the capsule. The amphicribral bundles are formed at the narrow base of the ovary by the fusion of vascular bundles and pass as placenta bundles through the placenta lobes running close to the inner ovary walls (Figure 6.19(A)).

A transverse section through the middle region of the ovary shows the wall with cuticularised epidermis and stomata. The sub-epidermal tissue consists of parenchymatous cells which are still vacuolating and have not differentiated into different layers of the pericarp. Placenta bundles run along the junction of the ovary wall and placenta which arise as voluminous lobes toward the free cavity in the center. The bundles show prominent laticiferous vessels which, instead of appearing as an arc in transverse section, appear in straight tangential rows (Figure 6.19(C), 6.20). Xylem, but not the phloem

FIGURE 6.18. Poppy capsules collected at weekly stages of development showing median vertical halves. W. 0–on the day of fall of petals; W. 1, 2, 3, 4–collected 1, 2, 3, and 4 weeks after fall of petals. (L. D. Kapoor)

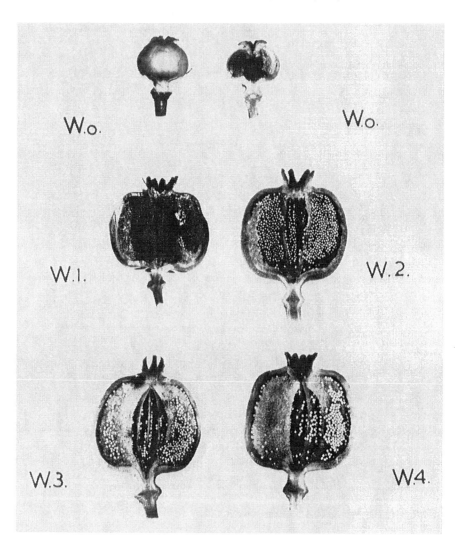

TABLE 6.4. Development of placenta bundle and laticifers in capsules

State of Development	Average diameter of placenta bundles (Microns)	Average diameter of laticifers in valve traces (Microns)
Week 0	750	5-20.5-27
Week 1	765	5-18.7-24
Week 2	855	15-18.6-24
Week 3	825	12-18.8-21

fibers, are lignified. The placenta bundle gives lateral branches towards the ovary wall (valves) where they give off sub-branches, but the density of these traces increases as the fruit develops.

In contrast to the general increase in the size of the fruit from ovary stage to Week 0 and then to Week 2, the diameters of laticiferous vessels do not show any significant increase as compared to the other tissues. Rather they are more prominent in the ovary stage than later and can be easily traced because of comparatively smaller cells of undifferentiated pericarp and other tissues. Some measurements of the diameters of laticifers in the placenta bundle and in the ovary wall are shown in Table 6.5.

The average diameter of laticifers in the central and peripheral vascular bundles of pedicel at this stage measured at 17.5-*18.9*-20.5μ and 14.3-*15.3*-16.8μ, respectively.

The process of articulation and anastomosing of laticifers can also be followed to some extent in the ovary wall. Two long narrow laticiferous cells which show nuclei embedded in thin granular cell contents lie end to end and their end wall is absorbed to form a single communicating cell. This process helps to increase the linear extension of the laticiferous vessel. But when two long narrow laticifers lie side by side, their adjacent lateral walls are absorbed. This may be absorbed totally and result in an increase in diameter of the laticifer. Sometimes, the absorption is partial or intermittent and results in the appearance of numerous oval orifices between the two cells (Figures 6.19(D), 6.21). The single long narrow cells show the presence of nucleus but when the contents of two cells fuse by the

FIGURE 6.19. Laticiferous vessels in ovary, Week (–1). A–L. S. of ovary showing valve and stigma traces. B–T. S. of ovary near the top and showing the
continuation of placenta with stigmatic rays, the three valve lobes are seen in an
oblique section at a little lower level. C–T. S. of ovary near the middle region.
D–laticifers in developing stages in the ovary wall. fl.tr.–floral traces; lat.–laticifers; ov.–ovules; ph.–phloem; pl.b.–placenta bundles; pl.p.–placenta plates;
pl.tr.–placenta traces; st.g.–stigmatic grooves; st.tr.–stigmatic traces; val.–
valves; val.tr.–valve traces. (L. D. Kapoor)

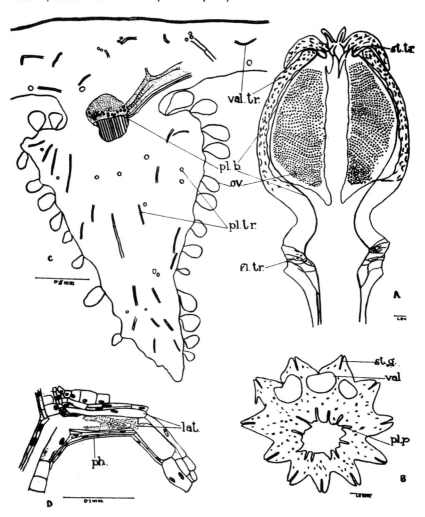

FIGURE 6.20. Photomicrograph of placenta bundle giving traces to ovary wall. lat.–laticifers; ph.–phloem; ph.f.–phloem fibers; xy.–xylem. X 280. (L. D. Kapoor)

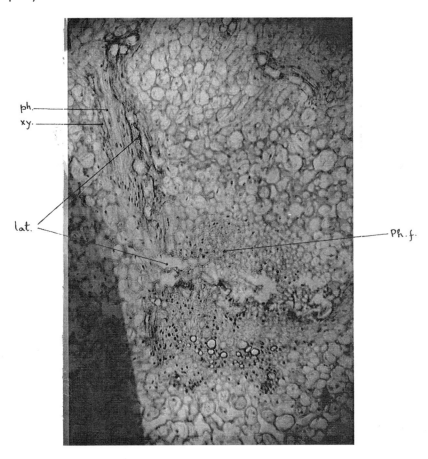

absorption of end walls or lateral walls the nucleus is obscured by the dense contents.

The placental bundles also give off traces toward the placenta where they ramify and supply the ovule. These traces are much smaller than those in the ovary wall. Their course and constitution is the same as has already been described in the placenta. Although these traces are much smaller in diameter than those in the ovary wall, their origin according to the order of development as observed by Bersillon[10] is earlier than those of the valve traces.

ANATOMY OF THE OTHER FLORAL PARTS

(a) *Sepals*

The sepals are pushed out and are abcissed when the flower bud opens. Each sepal has eight to 10 well-developed main veins which give off veinlets that ramify throughout the sepal. The venation of the sepals is a very complicated network, the branches of which show a very tortuous course.

In the transverse section of a sepal the main veins appear to be located centrally but the veinlets are concentrated more toward the abaxial side (Figure 6.22(S)). Each vein or veinlet shows the typical constitution of a vascular bundle consisting of an arc of xylem and phloem, the latter consisting mainly of phloem parenchyma and much less developed sieve tubes and companion cells (Figure 6.18(S.2)). Within the phloem, laticiferous vessels can easily be located appearing as a convex arc. In a longitudinal section articulated and anastomosing laticifers containing granular latex can be clearly detected running along the veins and following the course of the phloem when a veinlet is given off. In some sections, circular orifices formed in the process of articulating or anastomosing of laticifers could be observed (Figure 6.22(S.1)). The veins do not show any phloem fibers. The laticifers are restricted to the phloem and could not be seen in any interfascicular region of the sepals as reported by Leger.[7]

The mesophyll of the ground parenchyma consists of four to six layers of spongy parenchyma with isodiametric and loosely arranged cells in a lacunose tissue, and these cells are packed with

TABLE 6.5. Average diameter in microns of laticifers and sieve tubes of ovary, Week (−1)

	Placenta bundles		Valve traces	
	Laticifers	Sieve Tubes	Laticifers	Sieve Tubes
1.	23.7	13.6	23.1	10.4
2.	22.4	12.8	19.5	7.4
3.	22.7	13.0	21.2	9.5
4.	23.2	13.1	20.5	8.4
5.	24.0	13.8	21.5	9.6

chloroplasts. The epidermis on the abaxial and adaxial sides shows the development of stomata and deposition of the cuticle.

Laticifers in the vascular bundles of sepals can be seen in the sepal traces in the thalamus and their development into an anasto-mosing system can be followed in the young laticifers (Figure 6.14(D)). A young laticifer produces a small bulge on its lateral wall and a similar protrusion occurs on another laticifer which is running parallel. The intervening wall between the protrusions dis-integrates when they come into contact, and, finally, direct commu-nication between the two initials is established.

(b) *Petals*

The petals, which are snow-white, open generally during the day and fall off one or two days after they open. These are obtuse and the base is slightly thickened.

About seven to nine main veins enter the base and these run through the main body of the petals toward the obtuse apex and give off branches and sub-branches. The branching of veins and veinlets gives the appearance of reticulate venation.

The epidermis is covered with cuticle, and stomata are abundant. The mesophyll tissue consists of eight to nine layers of loosely arranged parenchymatous cells containing plastids. A similar tissue of five to six layers is seen below the vascular bundle on the abaxial side. The vascular bundle in transverse section shows the xylem consisting of tracheids with annular or helical thickening and xylem

FIGURE 6.21. Photomicrograph of laticifers showing development in ovary wall. lat.–laticifers–with nucleus; c.o.–circular orifices; e.w.–end walls showing absorption. (L. D. Kapoor)

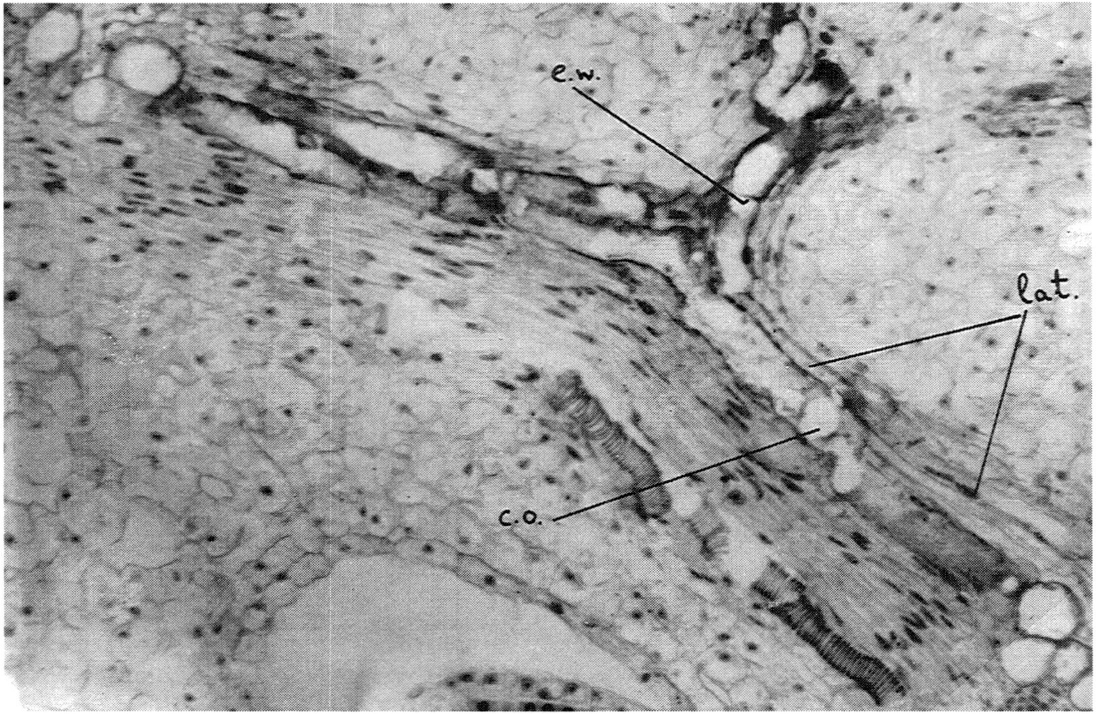

parenchyma (Figure 6.22.(P2)). Phloem consists mainly of phloem parenchyma. Sieve tubes do not appear to be well developed. Articulated laticifers are seen within the phloem and they follow the course of phloem tissue as a veinlet is given off from a vein (Figure 6.22(P.1)). A section through the marginal tip of the petal (Figure 6.22(P.3)), where a single file of tracheary element and phloem element is seen, indicates that the laticiferous vessels end before the vascular strands terminate. The traces for petals in the thalamus show the presence of laticifers, and their development can be followed in the same way as described for sepal traces.

(c) *Stamens*

There are numerous stamens, characteristic of the family, attached to the thalamus in two or three rows.

(i) *Filament*

In transverse and longitudinal sections the filament shows an epidermis with deposition of cuticle and presence of stomata. The ground tissue consists of vascuolated parenchymatous cells without intercellular spaces. The xylem consisting of a small lignified tracheids is surrounded by thin-walled cells of phloem elements (Figure 6.23.(F)). The presence of laticifers could not be confirmed. The specimens cleared in latic acid or macerated in dilute potassium hydroxide or sodium hydroxide did not reveal the presence of laticifers either. The position and orientation of vascular strand does not alter much when it enters the connective and terminates blindly near the apex of the anther. The vascular strand is not connected by any vascular element with the sporogenous tissue. The presence of a single vascular bundle in stamens confirms the findings of Wilson[13] and Eames,[14] who observed that in the majority of dicotyledons the stamens are supplied with a single vascular bundle which traverses the filament.

(ii) *Anthers*

Transverse and longitudinal sections of the anther show the cuticularized epidermis followed by highly specialized parenchymatous tissue which forms two to three layers of wall, (Figure 6.23(A.C.)), known as the fibrous layer and inner parietal layer of the microsporangia (pollen sac), but this tissue may be five to six layers toward the vascular bundle (Figure 6.23(B)). Fibrous layer cells are also seen scattered in the ground parenchyma near the parietal layer. The tapetum, which is the innermost parietal layer, is seen disintegrated and as a plasmodial mass concentrated at the base of microsporangium in the mature anther (Figure 6.23(B,D,E)). Xylem consists of tracheids, as in filament with annular secondary thickenings and thin-walled cells of phloem parenchyma found scattered around the xylem. Laticifers or phloem fibers have not been observed.

The absence of laticifers in stamens seems to be well confirmed, but the presence of laticifers in the stamen traces in the thalamus have been noticed. These traces apparently carry the laticifers, but it could not be confirmed where they terminate, as the epidermis along with sub-epidermil tissue is broken off where the stamens are attached to the thalamus.

(d) *Ovary*

The pistil of *Papaver somniferum* is described as poly carpellary syncarpous. The ovary is unilocular but partially divided by parietal placentation. The placenta, which are covered with ovules, do not reach the middle of the ovary. Morphologically the ovary is in the form of an urn resting upon a narrow base. A conical cap formed by the union of stigmatic rays tops the ovary.

(e) *Stigma*

The stigmatic rays in the young ovary present similar anatomy and bundle course as described for the mature stigmatic rays of the capsule. They are generally fused together by their edges and deeply grooved in the center. Papillae occur on their surface rather uniformly. The rays are in continuation with the placental plates and do not form a part of the ovary wall (Figure 6.19(A,B)). The placental bundle divides to supply the traces to stigma. Even in the young stage in these traces, laticifers can be seen and their development can be followed to some extent as described in the ovary wall.

It would thus be seen that the vascular supply to sepals, petals, and ovary all contain the laticifers associated with phloem. The stamens, however, do not appear to carry any laticiferous vessels, but the floral traces within the thalamus all show the presence of laticifers.

FIGURE 6.22. Laticiferous vessels in sepals and petals. S–T. S. of sepal near the base; S1–R. L. S. through a main vein; S2–T. S. of a vein in detail; P3–termination of a veinlet showing absence of laticifers; ep.–epidermis; lat.–laticiferous vessels; par.–parenchyma; ph–phloem; v.b.–vascular bundles; xy.–xylem element. (L. D. Kapoor)

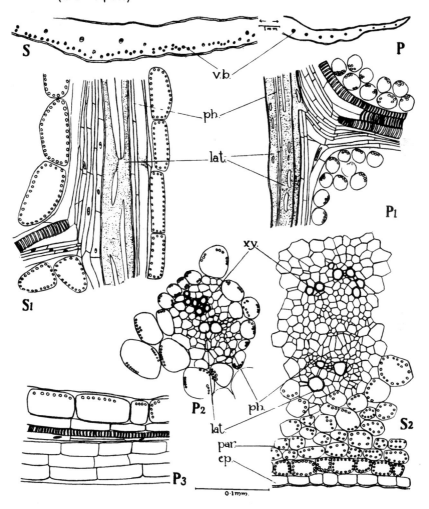

FIGURE 6.23. Histology of stamen showing absence of laticifers. A–diagram of L. S. of anther; B– details of connective at the position shown by arrow; C– details of outer epidermis of the anther; D– diagram of T. S. of another at the position (A-B) shown in A; E–details of T. S. of anther; F–details of T. S. of filament. ep–epidermis; i.p.l.–inner pariental layer; f.l.–fibrous layer; ph.–phloem; tap.–tapetum; xy.–xylem elements; A, D X 8; B, C, E, F X 280. (L. D. Kapoor)

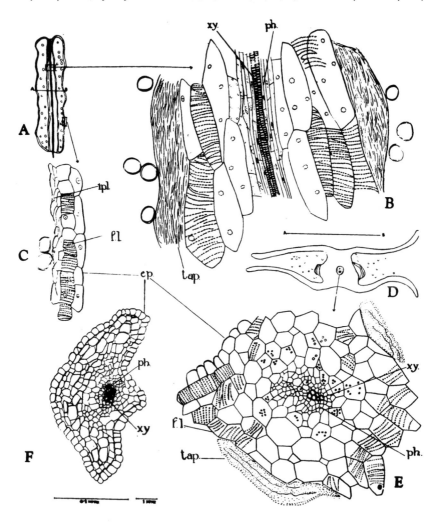

ANATOMICAL STUDIES OF SEEDS AND SEEDLINGS

(a) *Anatomy of seed with reference to the laticifers*

The seeds of opium poppy are very small, weighing about 35-45 mg to a hundred and consist of thin testa enclosing an oily endosperm in which a curved embryo is embedded. The cotyledons are slightly longer than the radicle. The procambial strands of the embryo could be observed in a sectioned material but no cells differentiating as laticifers could be identified. This is in contrast to the seeds of genus *Euphorbia* (which are comparatively much larger than poppy seeds) in which primoridial laticifers appear in the embryo when the cotyledons are initiated.[15]

(b) *The laticiferous vessels in the seedlings*

Seeds were germinated on a moist blotting paper and in sand culture. These sprouted in two to three days and showed typical epigeal growth. The cotyledons, which are small in size, somewhat fleshy, and linear to linear lanceolate in shape, opened about 12 days after sowing of the seeds. Within another two to three days the young foliar leaves protrude but they do not expand finally until another seven to eight days. It would thus appear that the cotyledons of opium poppy seeds at room temperature (19°C) opened in 12-13 days and the first foliage leaves expanded in 21-22 days after the sowing. The average length of the whole seedling at the cotyledonery stage was 1.0 cm and that at the stage when the first foliage leaves expanded was 2.5 cm.

The presence of laticifers in the radicle which had grown up to 5 mm in about seven to eight days could not be observed either in the cleared or in the sectioned material. It was not until the green cotyledons opened that the first appearance of laticifers could be noticed. When a cleared seedling was mounted in Chateliers reagent[16] and warmed gently, the developing laticifers showed the presence of minute yellowish-brown staining particles and these could be observed in close association with vascular traces of cotyledons, hypocotyle, and primary root.

In the cotyledons the laticifers are seen associated with phloem elements as the cotyledon trace passes into the lamina (Figure 6.24(A)). Their presence is similarly located and identified in the region of hypocotyle (Figure 6.24(B)). In the young primary root at this stage of the seedling these are found just above the region where root hairs are given off. They are much smaller in diameter than the upper parts of the seedling (Figure 6.24(D)).

In the early stages of development the laticifers appear as long narrow cells lying end to end or side by side with their end walls or lateral walls still intact. Nucleus can be seen embedded in the thin granular contents (Figure 6.24(A)). As the vascular traces grow, the end walls of laticifers are absorbed and their lateral walls (when they lie side by side) show intermittent disintegration with the result that the granular contents become dense, nucleus is obscured (Figure 6.24(B)), and laticifers elongate. The laticifers in their early stages of development also show circular or oval orifices but later on they are also obscured due to dense contents of latex. These orifices seem to develop due to intermittent disintegration of lateral walls when the contents of two cells tend to spread, leaving a gap or gaps where the walls have not disintegrated. The gaps show as orifices in longitudinal sections.

As the seedling develops the first foliage leaves, the laticifers become more prominent by their size and progressive articulations and anastomoses.

This could be seen in a longitudinal section of the main vein of the first foliage leaf (Figure 6.24(C)), which also shows the circular orifices in the laticifers.

It now appears that the laticifers which do not occur in the dry or sprouting seeds make their first appearance when the seedling has opened the cotyledons. They are simple long narrow cells containing a nucleus but later on the nucleus is obscured due to dense granular contents which follows the resorption of end walls of these cells. The intermittent absorption of lateral walls seems to cause the appearance of oval or circular orifices which are often seen in the laticifers. Thureson-Klein[17] observed that laticifers are not present in the embryos but differentiate soon after germination and are found in the phloem areas 18-30 hours after the seed is sown. Laticifers and sieve elements are generally separated by at least one cell layer in the roots but in colyedons, stems and leaves, they usually occur adjacent to each other.

As the laticifers differentiate an abundance of vesicles forms in the cytoplasm. This process appears to involve the endoplasmic

FIGURE 6.24. Laticifers in the seedlings. A–L. S. of cotyledon trace; B–L. S. of trace in hopocotyle; C–R. L. S. through main vein of first foliage leaf. D–L. S. of root above the "root hairs" zone. cort.–cortex; lat.–laticifers; ph.–phloem elements; xy.–xylem; A, B, C–same magnification. (L. D. Kapoor)

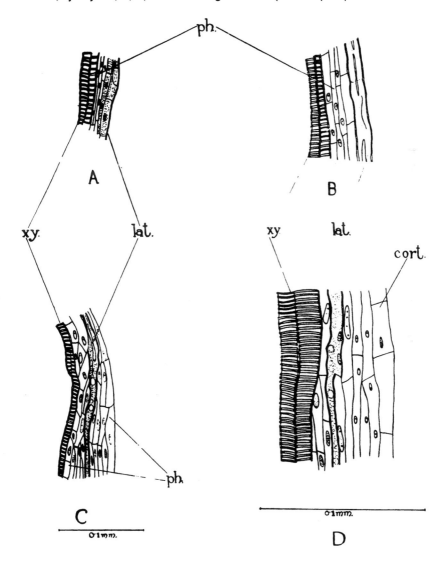

reticulum and it is suggested that the vesicles may be specialized forms of vacuole. Substances present in the vesicles react strongly with idine-potassium iodide. Laticifer-cytoplasm persists peripherally and between the vesicles. It contains the usual cell organelles, the presence of which substantiates an active metabolic role for the laticifer contents.

REFERENCES

1. Metcalf, C. R. and Chalk, L. Anatomy of Dicotyledons. Clarendon Press, Oxford, 1950.

2. Easau, K. Plant Anatomy. John Wiley and Sons, Inc., New York, 1953.

3. DeBary, A. Comparative Anatomy of the Phanerograms and Ferns. Clarendon Press, Oxford, 1884.

4. Frey-Wyssling, A. Die Sloffansscheidung der hoheren pflanzen. Monographien aus dem Gesmlgebiet der physiologie der pflanzen und der Tiere. Band 32, Julius Springer, Berlin, 1935.

5. Bonner, J. and Galston, A. W. Bot. Rev. 13, 1947, pp. 543-594.

6. Trecul, M. A. Comples. Rendus de l'Academie des Sciences. Tom 60, Jan.-June, R. 522, 1865.

7. Leger, J. Societe Linneenne de Normandi. Memoires, (18), 1984, p. 516.

8. Tschirch, A. and Oesterle, O. Anatomischer Atlas Der Pharmakognosie und Nahrungsmillelkunde. Leipzig Chr. Herm. Tauchnitz, 1900, p. 65.

9. Fedde, F. Papaveraceae in Engler and Prantls Die Naturl Pflanzenfam. Auf. 2, 17-B, 1936, p. 59.

10. Bersillon, G. Recherches sur les Papaveraceae. PhD Thesis. University of Paris, France, 1955.

11. Kapoor, L. D. The Laticiferous Vessels of *Papaver somniferum* L. PhD Thesis, London University, 1958.

12. Harvey-Gibson, R. J. and Bradley, H. M. Trans. Roy. Soc. Edinburgh. 51, 1917, pp. 589-608.

13. Wilson, C. L. The Telome Theory and the Origin of the Stamen. Amer. Jour. Bot. 29, 1942, p. 29.

14. Eames, A. J. The Vascular Anatomy of the Flower with Refutation of the Theory of Carpel Polymorphism. Amer. Jour. Bot. 18, 1931, pp. 147-188.

15. Cameron, D. Edinb. Trans. and proc. Bot. soc. 32.1, 1936, pp. 187-194.

16. Rowson, J. M. Mountants for use in Pharmacognosy. Pharm. J. 165, 1950, p. 60.

17. Asa Thureson-Klein. Observations on Development and Fine Structures of the Articulated Laticifer *Papaver Somniferm*. Ann. Bot., 34, 1970, pp. 751-775.

Chapter 7

Constitution of Amphicribral Vascular Bundles in Capsule

Amphicribral concentric vascular bundles are frequently encountered in stems of ferns, where they vary considerably in transverse sectional shape.[1] De Bary observed that amphicribral vascular bundles are apparently rare in angiosperms.[2] Their occurrence in diocotyledons has been reported in flowers, fruit, ovules, and leaves (Eames and MacDaniels),[3] stamens (Leinfellner),[4] and pith.[2]

While studying the distribution and anastomosing of laticifers in *Papaver somniferum*, amphicribral bundles were observed in the receptacle and capsule. Transverse sections in acropetal succession revealed their formation and dissolution.

VASCULAR BUNDLES OF THE PEDICEL

A transverse section of the pedicel at its distal end reveals numerous vascular bundles. There is an inner ring of nine to 13 large vascular bundles and an outer peripheral zone of 39 to 41 moderately scattered vascular bundles (Figure 6.3 (1,2). All the bundles are collateral (Figure 6.7). Each central vascular bundle alternates with approximately three vascular bundles, the middle one of which is the largest. The number of vascular bundles varies as indicated, even within the same population. The variation in central and peripheral bundles is correlated so that the association of one central bundle with three peripheral ones is fairly constant.

The peripheral ground tissues of the pedicel are composed of an inner five to six layers of sclerotic, pitted, lignified parenchyma hardly distinguishable from the phloem fibers and an outer five to seven layers of chlorenchyma with prominent intercellular spaces. The sclerotic layer is confluent with the protophloem fibers in all but a few central bundles. These tissues are bound by a hypodermis and an epidermis with a thick cuticle. The central ground tissue among the bundles is sclerotic.[5]

The protophloem fibers are narrow, longer than the lignified parenchyma, and have small simple pits. The functional primary phloem consists of phloem parenchyma sieve tubes, companion cells, and laticiferous vessels. The laticiferous vessels are of the articulated type and anastomose among themselves within a bundle but do not anastomose with those of adjacent bundles. The laticiferous vessels form a discontinuous convex arc in the peripheral bundles (Figure 6.7), and a concave discontinuous arc in the central bundles (Figure 6.8). In the latter, the arc of laticiferous vessels may almost surround the protophloem fibers. The arc of laticiferous vessels frequently occurs near the boundaries of primary and secondary phloem. Inside the phloem are, successively, the cambium and xylem. The xylem consists of tracheids with annular or spiral thickenings and xylem parenchyma.

THE NATURE AND COURSE OF THE VASCULAR BUNDLES IN THE RECEPTACLE AND GYNOECIUM

In the receptacle[8] the course of vascular bundles as seen in the pedicel (Figure 6.3(1-3)) remains unchanged until the region of the receptacle is reached (Figure 6.2(B)). The bundles in this region give rise to traces which may anastomose and branch again. Many of these traces run vertically for very short distances but soon bend toward the cortex and supply the perianth members (Figures 6.3(4), 6.4(5)). The traces to sepals and petals consist of xylem, phloem, and laticifers, but not phloem fibers. As far as could be determined, the bundles or traces always anastomose before giving rise to new branches. Hence, in transverse section the number of bundles seen may increase or decrease as the receptacle is ascended. In ascending serial sections, the sequence appears as follows: Some of the outer bundles join the inner ring of large bundles to form a central ring of about 26 vascular bundles. The remaining outer 26 bundles anastomose by pairs, and then each gives rise to two traces which presumably supply the sepals and petals (Figures 6.4(6,7)). Higher, the large inner bundles divide once more to form a peripheral ring of about 50 equal-sized bundles (Figure 6.4(8)). Three or four of these bundles then fuse to form 11 amphicribral bundles which supply the base of the capsule (Figure 6.15). The chlorenchyma tissue ends in the region where most of the stamen traces are beginning (Figure 6.4(6,7)) and then reappears when all the stamen traces have been given off (Figure 6.4(8)).

In the process of fusion, the xylem assumes a reniform shape in transverse section. The following occurs in acropetal succession: In each bundle the phloem extends around one side of the xylary strand (Figure 6.4(9)). Two adjacent vascular bundles with their xylary strands facing each other join. Xylem joins xylem and the phloem becomes continuous around the resulting xylem strand to form an amphicribral bundle (Figure 6.5(12)). The bundle is surrounded by phloem fibers (Figure 6.15).

In the capsule the number of amphicribral bundles varies with the plant, but the number entering the capsule is generally equal to the number of placentae and stigmatic rays. They are the principal bundles of the capsule and lie just to the outside of the placentae and opposite the longitudinal grooves in the surface of the capsule. Each bundle, called a placental bundle,[6] apparently represents the product of two ventral carpellary veins, each formerly belonging to different adjacent carpels; these bundles give rise to ovular bundles which supply the many ovules (Figure 6.5(13), Figure 6.6(14,15)). No dorsal carpellary bundles are evident, and the valves (dorsal walls) are supplied by many small branches from the placental bundles (Figure 6.5(10,11,12,13), Figure 6.6(14,15)). Near the top of the capsule where the placental bundles bend toward the stigma, they become progressively reduced from an amphicribral to a collateral type of vascular bundle (Figure 6.6(14,15,16,17,18)).

It would thus appear that the placental bundles form the main vascular framework of the capsule. The valve bundles, which run in the capsule wall and anastomose freely, arise as branches from the placental bundles (Figure 6.5(10,11,12,13)). The vascular supply to the placentae and ovules also arises from the placental bundles (Figure 6.6(14)), as do vascular bundles in the stigmatic rays.

In the stigmatic rays, the intercarpellary regions are progressively reduced and finally disappear as the placental bundles reach the distal end of the capsule. The stigmatic rays are situated above the placentae. Each distal collateral derivative of an amphicribral placental bundle bifurcates above its placenta, and each branch fuses with a comparable adjacent bundle. The resulting product supplies the stigmatic rays with numerous small bundles. The vascular bundles in the stigmatic rays carry the laticiferous vessels associated with poorly developed phloem (Figure 6.6(14,15,16,17, 18)).

The phloem fibers which form characteristic caps on the vascular bundles were not observed on the bundles supplying the ovules, valves, or stigmatic rays. The laticiferous vessels are present in all the latter bundles, but they do not extend into the ovules.

REFERENCES

1. Russow, E. Vergleichende Untersuchungen den Leit Bundle-Kryptogamen. Mem. Acad. Imp. Sci. Saint Petersbourg, ser. 7, 19, 1872, pp. 1-207.

2. De Bary, A. Comparative Anatomy of Vegetative Organs of Phanerogams and Ferns. Clarendon, Oxford, 1884.

3. Eames, A. J., and Mac Daniels, L. H. An Introduction to Plant Anatomy. McGraw-Hill, New York, 1947.

4. Leinfellner, W. Die Gefass Bundelversorgung des Liliumslaubblattes, Osterreichische Bot. Z, 103, 1956, pp. 346-352.

5. Harvey-Gibson, R. J. and Bradley, H. M. Trans. Roy. Soc. Edinburgh, 51, 1917, pp. 589-608.

6. Bersillon, G. Recherches sur les Papaveracees. Thesis, University of Paris, 1955.

7. Johansen, D. A. Plant Microtechnique. McGraw-Hill, New York, 1940.

8. Kapoor, L. D. The Laticiferous Vessels of *Papaver somniferum*. PhD thesis. London University, 1958.

9. Foster, A. Practical Plant Anatomy, Van Nostrand, New York, 1950.

10. Horne, R. L. Van, and Zoff, L. C. Water Soluble Embedding Materials for Botanical Micro-technique, J. Amer. Pharm. Ass., 40, 1951, p. 31.

11. Kapoor, L. D. Constitution of Amphicribal Vascular Bundles in Capsule of *Papaver somniferum*, Bot. Gaz, 134(3), 1973, pp. 161-165.

Chapter 8

Chemical and Pharmacological Aspects of Opium Alkaloids

Trease and Evans[1] described that opium (raw opium) of B. P. (British Pharmacopea) is the latex obtained by incision from the unripe capsules of *Papaver somniferum* L. and dried partly by spontaneous evaporation and partly by artificial heat. It is worked into irregularly shaped masses and is known in commerce as Indian opium; raw opium contains no less than 9.5 percent morphine. Indian opium is specifically stated because this is now the only legally available source of the drug. However, a number of countries (e.g., Turkey, Pakistan, the former USSR, the former Yugoslavia, Tasmania, etc.) grow considerable quantities of the opium poppy for alkaloid extraction and seed production. Considerable quantities of illegal opium are produced in Southeast Asia.

The characteristics of opium B. P. are the presence of a water insoluble portion consisting of epidermal fragments of the capsule showing in surface view, five to six sided, thick-walled cells and large anomocytic stomata; fragments of poppy leaf, the lower epidermis of which possesses slightly wavy walls and anomocytic stomata.

POWDERED OPIUM

Official powdered opium is standardized to contain 10 percent anhydrous morphine. Its strength may be adjusted with lactose colored with burnt sugar or with powdered cocoa husk. The microscopic examination shows the characteristic sclereids and mucilage cells of the cocoa as well as the amorphous latex masses and leaf fragments of the opium poppy. It exhibits brown stone cells, narrow spiral vessels, and parenchyma. For powdered opium the B. P. directs that the raw opium should be dried at moderate temperature and reduced to a moderately fine powder; it is adjusted to strength as described above.

Papaveretum or *Opium Concentratum* is a standard preparation of the hydrochlorides of the opium alkaloids containing 47.5-52.5 percent anhydrous morphine, 2.5-5 percent of codeine, 16-22 percent noscopine (narcotine), and 2.5-7.0 percent papaverine.

Opium contains more than 40 alkaloids which are largely combined with the organic acid, meconic acid; the drug also contains sugars, salts (e.g., sulphates), albuminous substances, coloring matter, and water. The following list summarizes some of the facts about the chief opium alkaloids:

Alkaloid	Formula	Discoverer	Date	Properties
Morphine	$C_{17}H_{19}O_3N$	Sertürner	1816	Strong bases, which are alkaline to litmus and highly toxic.
Codeine	$C_{18}H_{21}O_3N$	Robiquet	1832	
Thebaine	$C_{19}H_{21}O_3N$	Thiboumery	1835	
Noscapine	$C_{22}H_{23}O_7N$	Derosne	1803	Feeble bases, which are only slightly toxic.
Narceine	$C_{23}H_{22}O_8N$	Pelletier	1832	
Papaverine	$C_{22}H_{21}O_4N$	Merck	1848	

The first group (e.g., morphine) consists of alkaloids which have a phenanthrene nucleus. Those of the Papaverine group have a benzylisoquinoline structure. Some of the less important opium alkaloids (e.g., protopine and hydrocotarnine) are of different structural types.[1] The morphine molecule has both a phenolic and an alcoholic hydroxyl group and when acetylated forms diacetyl morphine or heroin. Codeine is an ether of morphine (methyl morphine). Other morphine ethers which are used medicinally are ethylmorphine and pholcodine.

MECONIC ACID

Meconic acid is easily detected either in the free state or as a meconate by the formation of deep red color upon addition of a solution of ferric chloride. As it is invariably found in opium, its presence has been long used to indicate opium.[1]

Its presence has been reported in sections Orthorhoeades, Argemonorhoeades, Mecones, Miltantha, Pilosa, Oxytona (Macrantha), and Scapiflora of *Papaver*. Fairbairn and Williamson[2] investigated its distribution in *Papaver* and related genera of Papaveraceae. Of 48 species of Papaveraceae representing 12 genera, 24 were shown to contain meconic acid and of these 21 were *Papaver* species. *Meconopsis cambrica* was also found to contain meconic acid in contrast to other species of *Meconopsis*. It was found in two *Roemaria* species also. It was suggested that the presence of meconic acid may be a distinctive chemo-taxonomic feature of *Papaver* and closely related genera.

Meconic acid

Source: Fairbairn and Williamson, 1978.

Meconic acid, a dibasic acid ($C_7H_4O_7$), has an unusually low H content in relation to C and O. It seems also to be restricted to *Papaver* species which characteristically produce morphinanes. Fairbairn and Steele[3] have investigated the possibility that it is an energy-exhausted product of the citric acid cycle whose formation controls alkaloid production. The concentration of meconic acid and of alkaloids in the latex have been shown to vary significantly during the day which suggests active metabolism but lacks firm evidence of formation of meconic acid from either fumarate or pyruvate.

Henry[4] observed that in 1803 Derosne, an apothecary practicing in Paris, observed the separation of a crystalline substance when a syrupy aqueous extract of opium was diluted with water. This crystalline material was probably narcotine or a mixture of that alkaloid with morphine. Seguin in 1804 read to the Institute of France a paper in which he described the isolation of morphine although he did not recognize its basic character. This paper was not published until 1814 and in the meantime Sertürner had obtained both morphine and meconic acid from opium and pointed out that the morphine was the first member of a new class of substances, "the vegetable alkalis." The composition of the alkaloid was first determined by Liebig in 1831, who represented it by the formula $C_{34}H_{36}O_6N_2$, which was reduced by Laurent in 1847 to the simple formula $C_{17}H_{19}O_3N$ now in use. However, the structures of these alkaloids were eventually deduced by Gulland and Robinson[5,6] after considerable experimental efforts.

A list of important alkaloids compiled by Thakur[7] along with the authorities who isolated or studied them is provided in Table 8.1.

On the average, a good sample of opium should contain 50 percent water soluble portion with ash content not more than 4 percent The major alkaloids normally assayed in opium are given below[8]:

Morphine	8.0	– 14.0%
Codeine	2.5	– 3.5%
Narcotine	1.0	– 7.5%
Papaverine	0.50	– 1.00%
Thebaine	0.10	– 2.0%

Santavy,[9] in a review of Papaveraceae alkaloid in 1970, listed the following alkaloids and the intermediates detected in *Papaver somniferum* of section Mecones:

Codamine, codeine, codeinone, coptisine, corytuberine, cryptopine, N-methyl-14-0-desmethylepiporphyroxine, glaudine, gnoscopine, hydrocotarnine, 10-hydroxycodeine, (+)-isoboldine, (-)-isocorypalmine, lanthopine, laudanidine, laudanine, laudanosine, magnoflorine, morphine, narceine, narcotine, narcotoline, neopine, nornarceine, papaveraldine, papaveramine, papaverine, papaverrubine C, porphyroxine (papaverrubine D), protopine, pseudomorphine, (±) -reticuline, salutaridine, salutaridinol-I, sanguinarine, (-)-scoulerine, thebaine, base X, allocryptopine, berberine, palaudine

In 1979 Santavy[10] added another list of alkaloids and their intermediates, isolated, or reinvestigated, or those for which the already isolated alkaloids have been examined for structural elucidation.

6-Acteonyldihydrosanguinarine, B-allocryptopine, bases mp 147°, 257°, and 260°, "Bound-morphine," canadine, choline, codeine, coreximine, cryptopine, dihydroprotopine, dihydrosanguinarine, gnoscopine, 16-hydroxythebaine, magnoflorine, 6-methylcodeine, narceine imide, narcotine, normorphine, norsanguinarine (callus tissues), orientaline, 13-oxocryptopine, oxydimorphine, oxysanguinarine, pacodine, palaudine, papaveraldine, papaverine, salutaridine, sanguinarine, stepholidine, thebaine, tetrahydropapaverine, two N-oxides of morphine, codeine, and thebaine.

Brochmann-Hanssen[11] observed that 35 or more opium alkaloids which have been isolated to date belong to eight major groups as illustrated by their carbon skeletons in Figure 8.1. Each group is represented in the plant by several members having different substitution patterns and oxidation states.

Bentley[12] stated that the alkaloids of the morphine group form a subgroup of the isoquinoline alkaloids being derived in nature from the bases of the laudanosine series by oxidative processes. Their pharmacological and chemical properties are sufficiently distinct to warrant their separate discussion. Morphine, codeine, neopine, thebaine, and oripavine, all isolated from *Papaver* species, form one of the enantiomorphic sub-groups; the other sub-group covers sinomenine, hasubanonine, metaphenine, and protometaphenine, all isolated from Japanese plants of the *Sinomenium* and *Stephania* species (Figure 8.2).

These bases are all closely related. Morphine ($C_{17}H_{19}NO_3$) is a phenol and gives codeine ($C_{18}H_{21}NO_3$) upon methylation. Neo-

TABLE 8.1. Important alkaloids of opium poppy

S1. No.	Name and Type	Percentage Occurrence	Authorities
	A. MORPHINANE		
1.	Morphine	3.00-25.00	Emde (1930), Hoffman la Roche (1937), Gates and Tschudi (1952), Achor & Geiling (1954), McGurie et al. (1957).
2.	Codeine	0.5-4.0	Gulland & Robinson (1926), Shaposhnikov (1937), Goto and Yamamoto (1954).
3.	Thebaine	0.4	Mannich (1916), Gulland and Robinson (1925), Schopf and Winterhalder (1927), Small & Browning (1939), Ghosh & Robinson (1944), Smith et al. (1973).
4.	Pseudo-morphine	—	Wu & Dobberstein (1977).
5.	Neopine	0.005	Gulland & Robinson (1923), Dubin, Robinson, and Smith (1926).
6.	Codeinone	Tr	Findlay & Small (1950).
7.	10-Hydroxycodeine	Tr	Dobbie & Lauder (1911).
8.	Salutaridine	Tr	Barton et al. (1965), Stuart (1971), Kametani et al. (1972).
9.	Salutaridinol	Tr	Santavy (1970).

Sl. No.	Name and Type	Percentage Occurrence	Authorities
B. BENZYLISOQUINOLINE			
10.	Papaverine	0.50-1.00	Pictet & Gams (1909), Rosenmund, Nothnagal, & Rissenreldt (1927), Mannich & Walther (1927), Spath & Burger (1927), Kinder & Peschke (1934), Gurthrie, Frank, & Purves (1955).
11.	Laudanine	0.01	Spath (1920), Spath & Lang (1921).
12.	Laudanidine	0.003	Spath & Seka (1925), Spath & Bernauner (1925), Ferrari & Deulofeu (1962).
13.	Codamine	0.001	Spath & Esptein (1926), Schopf & Thierfelder (1939), Billek (1956).
14.	Laudanosine	Tr	Pictet & Finkestein (1909), Craig & Tarbell (1948), Kondo & Mori (1931). (1944).
15.	Papaveraldine	Tr	Dobson & Perkin (1911), Buck, Haworth, & Perkin (1924), Menon
16.	Reticuline	Tr	Gopinath et al. (1959), Brochmann-Hanssen, & Furaya (1964).
17.	Palaudine	Tr	Santavy (1970).

TABLE 8.1 (continued)

S1. No.	Name and Type	Percentage Occurrence	Authorities
	C. APORPHINE		
18.	Corytuberine	Tr	Dobbie & Lauder (1893).
19.	Isoboldine	Tr	Santavy (1970).
20.	Magnoflorine	Tr	Gopinath et al. (1959).
	D. PROTOBERBERINE AND TETRAHYDROBERBERINE		
21.	Coptisine	Tr	Spath & Posega (1929), Klasek, Simanek, & Santavy (1968).
22.	Isocorypalmine	Tr	Spath & Burger (1926), Govindachari, Rajadurai, & Ramadas (1959).
23.	Scoulerine	Tr	Santavy (1970).
24.	Berberine	Tr	Santavy (1970).
	E. PROTOPINE		
25.	Cryptopine	Tr	Perkin (1916), Perkin (1919), Haworth & Perkin (1926), Anet & Marion (1954).
26.	Protopine	Tr	Perkin (1916), Haworth & Perkin (1926).
27.	Allocryptopine	Tr	Bentley & Murray (1963).

S1. No.	Name and Type	Percentage Occurrence	Authorities
	F. PHTHALIDEISOQUINOLINE		
28.	Narcotine	1.0-12.0	Perkin & Robinson (1911), Polonovski & Polonovski (1930), Marshall, Pyma, & Robinson (1934), Barnes (1955) Ohta et al. (1963).
29.	Narcotoline	Tr	Wrede (1937), Baumgarten, & Christ (1950), Pfeifer & Weiss (1955), Pfeifer (1957).
30.	Narceine	0.2	Addinall & Major (1933).
31.	Nor-narceine	Tr	Santavy (1970).
32.	Rhoeadine	Tr	Nameckova & Santavy (1962), Hiroshi, Shohei, & Yamane (1972).
33.	Gnoscopine	Tr	Rabe & McMillan (1910), Marshall, Pyma, & Robinson (1934).
	G. RHOEADINE		
34.	Glaudine	Tr	Santavy (1970).
35.	Papaverrubine-C	Tr	Santavy (1970).
36.	Papaverrubine-D	Tr	Santavy (1970).
	H. BENZOPHENANTHRIDINE		
37.	Sanguinarine	Tr	Spath & Kuffner (1931), Dyke, Moon, & Sainsbury (1968), Hakim, Mijovic, Walker (1961).
	I. TETRAHYDROISOQUINOLINE		
38.	Hydrocotarinine	Tr	Topchiev (1933).

FIGURE 8.1. The carbon skeletons of opium alkaloids (E. Brochmann-Hanssen). Courtesy of Springer-Verlag Publishers, Heidelberg, 1971.

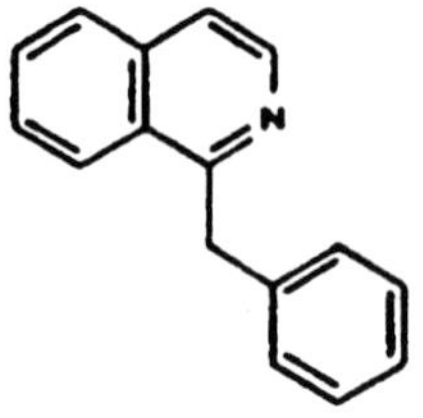

BENZYLISOQUINOLINE 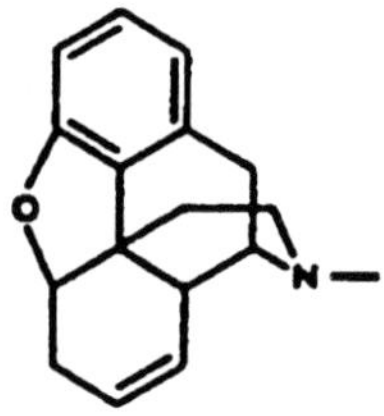   HYDROPHENANTHRENE 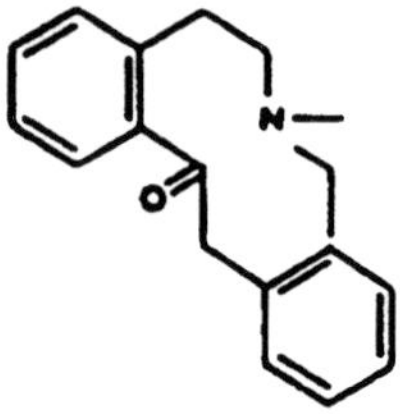   PROTOPINE

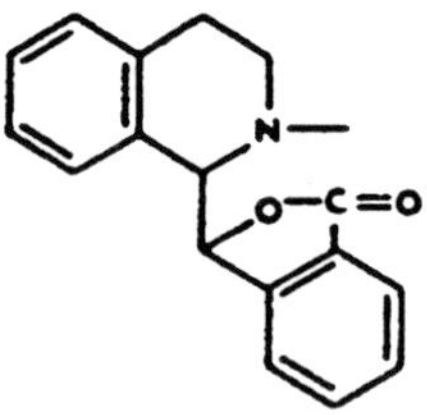

PHTHALIDE · ISOQUINOLINE 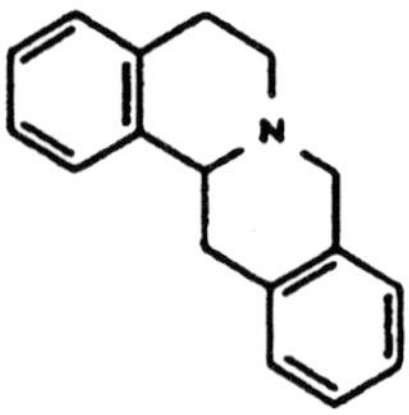   PROTOBERBERINE 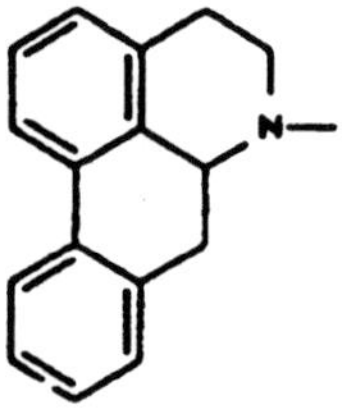   APORPHINE

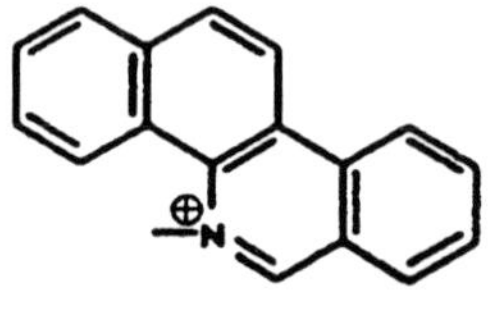

BENZOPHENANTHRIDINE 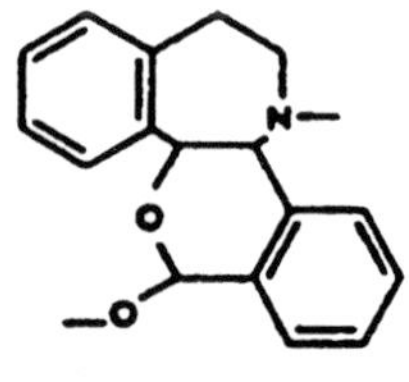   PAPAVERRUBINE

pine, which is isomeric with codeine, gives dihydrocodeine upon catalytic reduction and can only differ from codeine in the position of a double bond. Codeine is also a secondary alcohol and can be oxidised to an a, B- unsaturated ketone, codeinone ($C_{18}H_{19}NO_3$) obtainable in poor yield by the careful hydrolysis of thebaine ($C_{19}H_{21}NO_3$) which must therefore be an enol methyl ether of

FIGURE 8.2. Alkaloids of opium (enantiomorphic subgroup) (Santavy, 1970). Courtesy of Academic Press, Inc.

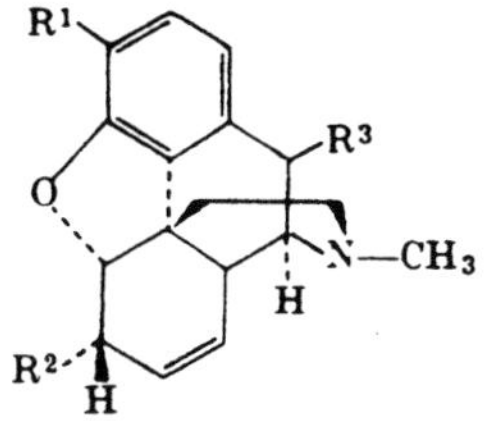

Morphine	R^1 = OH, R^2 = OH, R^3 = H
Codeine	R^1 = OCH$_3$, R^2 = OH, R^3 = H
10-Hydroxycodeine	R^1 = OCH$_3$, R^2 and R^3 = OH
Codeinone	R^1 = OCH$_3$, R^2 + H = O, R^3 = H

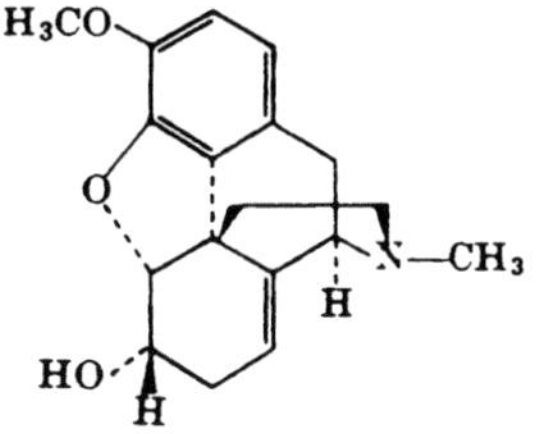

Neopine

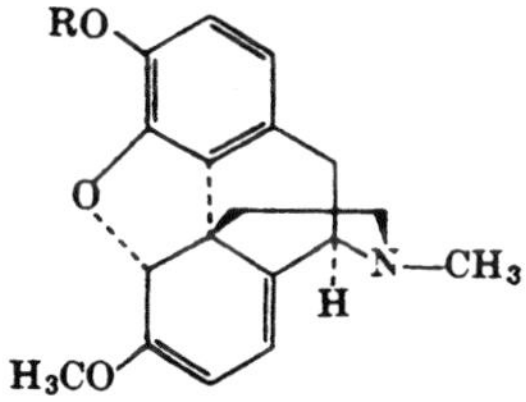

Oripavine R = H
Thebaine R = CH$_3$

codeinone. Oripavine, like morphine, is a phenol and on methylation with diazomethane adds CH$_2$ to give thebaine.[12]

Cordell[13] observed that over 40 alkaloids are known and several are of major significance. These are morphine, codeine, thebaine, noscapine, and papaverine. The morphine content of opium is in the range 4-21 percent, of noscapine 4-8 percent and the others in the range 0.5-2.5 percent. These alkaloids occur at least partially bound to meconic acid, the presence of which can also be used to detect opium.

Some of these alkaloids used medicinally are discussed along with their pharmacology and uses.

Santavy[10] reported the additional compounds, viz. 6-methyle codeine, 16-hydroxythebaine, normorphine, 14B-hydroxycodeine, and 14B-hydroxycodeinone as new alkaloids of the morphinane type. Normorphine was found in opium with the help of sensitive reactions and it was established as an active metabolite of morphine in *P. somniferum*. In the *Papaver* species there were always demonstrated two isomeric N-oxides of codeine, morphine, and thebaine which were also prepared synthetically (Figure 8.3).

The new alkaloid oripavidine isolated from *P. orientale* was identified as 3,13-didemethylthebaine.

By using ^{14}C-morphine it could be shown that morphine is further degraded to such non-alkaloidal metabolites as thebaol, which is elaborated directly by the plant from thebaine[10] (Figure 8.4).

FIGURE 8.3. Additional compounds of opium alkaloids (Santavy, 1979). Courtesy of Academic Press, Inc.

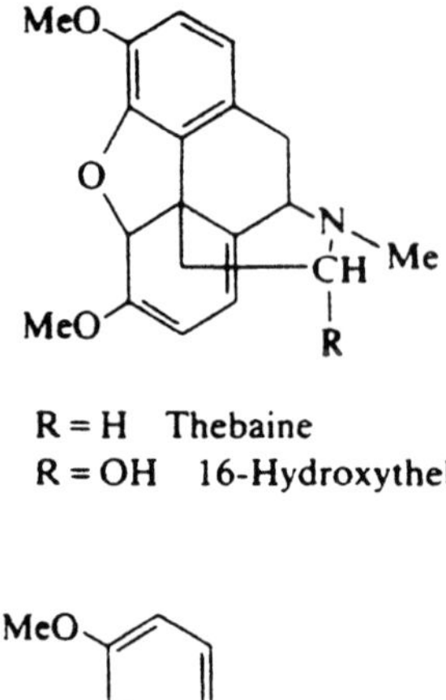

R = H   Thebaine
R = OH   16-Hydroxythebaine

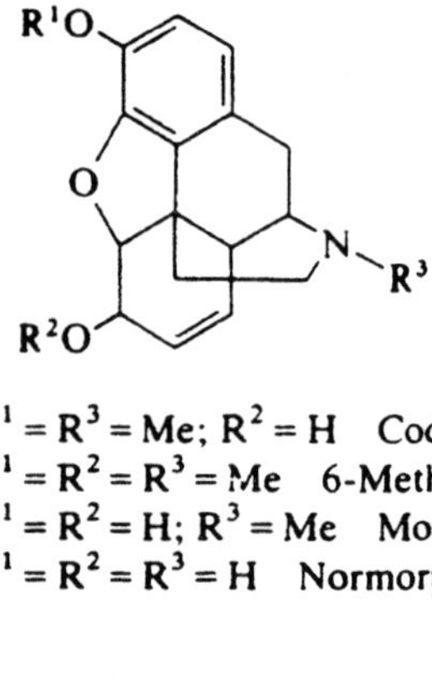

$R^1 = R^3 = Me; R^2 = H$ Codeine
$R^1 = R^2 = R^3 = Me$ 6-Methylcodeine
$R^1 = R^2 = H; R^3 = Me$ Morphine
$R^1 = R^2 = R^3 = H$ Normorphine

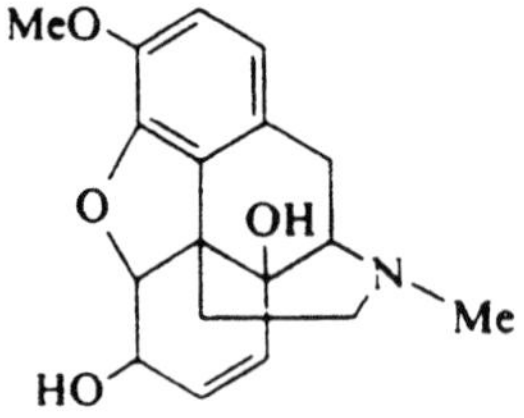

14-Hydroxycodeine

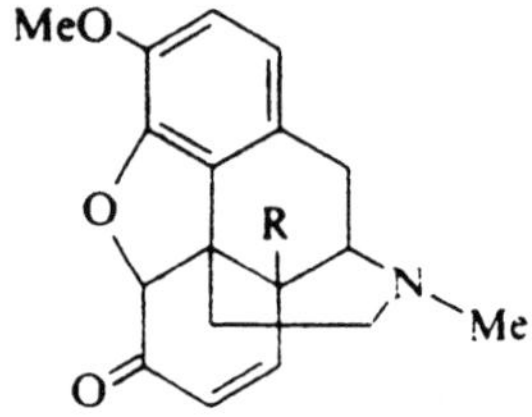

R = H   Codeinone
R = Br   14-Bromcodeinone
R = OH   14-Hydroxycodeinone

PHARMACOLOGY

Cordell[13] reported that the pharmological actions of morphine are complex. Some specific central nervous system functions are depressed while others are stimulated. There is also some stimulation of sympathomimetic and parasympathomimetic systems.

Typically, morphine produces analgesia, drowsiness, changes in mood, and mental clouding. A significant feature of morphine-induced analgesia is that it occurs without loss of consciousness. There are two factors involved in the relief of pain resulting from the analgesics; they elevate the pain threshhold and they alter the reaction of the individual to the painful experience. Drowsiness commonly occurs, the extremities feel heavy, and the body feels warm. Some patients also experience euphoria. The desire to create

FIGURE 8.4. Compounds of opium alkaloids (Santavy, 1979). Courtesy of Academic Press, Inc.

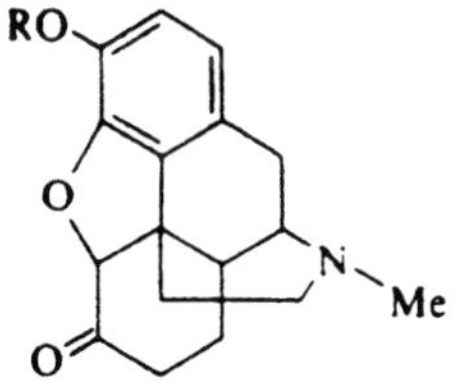

R = Me   Dihydrocodeinone
R = H   Dihydromorphinone

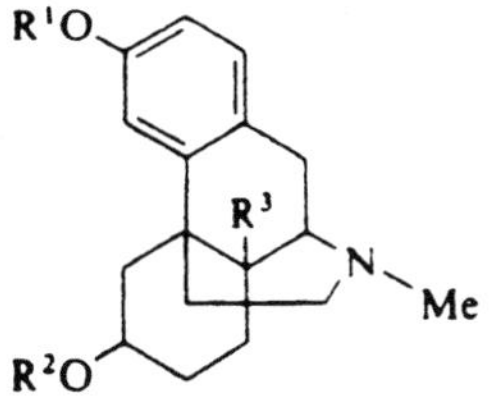

R^1 = Me; R^2 = R^3 = H   Dihydrocodeine
R^1 = Me; R^2 = H; R^3 = OH   14-Hydroxydihydrocodeine
R^1 = R^2 = R^3 = H   Dihydromorphine

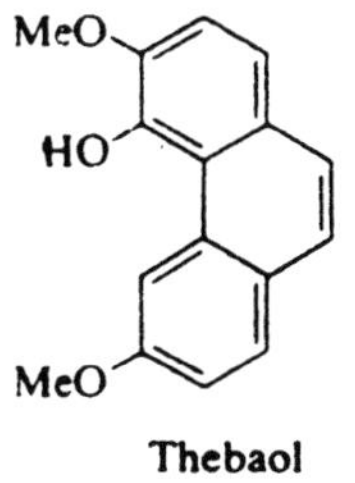

Thebaol

this feeling of euphoria is purportedly the major cause of dependence.

When opiates are given to a patient who has not previously experienced the effects of the drug and who is not in pain, there is a sense of attachment and a feeling of anxiety or uneasiness/dysphoria. In this dysphoric state, nervousness, fear, nausea, and vomiting may occur. In humans, death from morphine poisoning is nearly always due to respiratory arrest. Therapeutic doses of morphine in humans depress all phases of respiratory activity: respiratory rate, minute volume, and tidal exchanges. The diminished respiratory volume is caused by a decrease in respiratory rate. Breathing may also be irregular and periodic.

As tolerance develops to the analgetic and euphoric effects, the respiratory center also becomes tolerant. This is the reason the addicts may exhibit resistance to otherwise lethal doses of morphine. Therapeutic doses of morphine and related narcotic analgesics have no major effect on blood pressure or heart rate and rhythm in the normal recumbent patient.

Morphine has a synergistic effect with other respiratory depressants such as other opiates, barbiturates, general anesthetics, and alcohol. The depressant effects of some opioids may themselves be exaggerated by the phenothiazines, monoamine oxidase inhibitors, and tricyclic antidepressants.

The opiates decrease the HCL secretion in the stomach and also cause a decrease in motility. There is an increase in tone of the first part of the duodenum, which delays passage of the gastric contents for as much as 12 hours. This is probably the main reason for the constipating effects of the morphine-like drugs. The second major effect is that the propulsive peristaltic contractions in the large intestine are decreased. Cordall further observed that in recognizing morphine intoxication, the decrease in pupillary size is of considerable practical importance. The consensus is that this action is centrally mediated.

Morphine and its analogues are powerful depressants of the cough center and reduce the awareness of coughing. Other actions of therapeutic doses include dilation of cutaneous blood vessels. This action is thought to be due to histamine release, and this may be responsible for the itching and sweating that commonly ensues following administration of morphine.

CODEINE

Codeine was first isolated from opium by Robiquet in 1833, when it was closely associated with morphine in concentration range 0.7-2.5 percent. For pharmaceutical requirements codeine is commercially produced by methylation of morphine as its production is insufficient by isolation.[13]

Cordell[13] observed that codeine is a very important analgetic and antitussive drug. It is less sedative and analgetic than morphine and tolerance develops more slowly than morphine. It has less effect on the gastrointestinal tract and on the genitourinary tract than morphine. In therapeutic doses it depresses the respiratory system only slightly, but toxic doses may produce respiratory stimulation with excitement and convulsions.

HEROIN

Heroin (diacetyl morphine) is a highly euphoric and analgetic drug. It crosses the blood-brain barrier more rapidly than morphine and is preferred by addicts because of the initial orgiastic sensation. The euphoria produced by heroin is greater than that of morphine. Heroin is approximately three to five times more potent than morphine as an analgesic and is also more potent in suppressing the cough center and in causing respiratory depression.

Hydrocodone and 6, 7-Dihydromorphine

The pharmacologic actions of hydrocodone (dihydrocodeinone) are midway between codeine and morphine. It shows a lower incidence of side effects and may be more effective than codeine as an antitussive. Unfortunately, it may also be more addictive than codeine (Figure 8.5).

As an analgesic, 6-7-Dihydromorphine is six to ten times more

FIGURE 8.5. (G. A. Cordell, 1981). Courtesy of John Wiley & Sons, Inc.

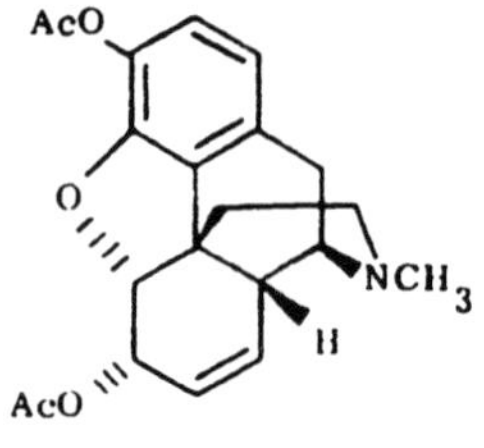

diacetylmorphine
(heroin)

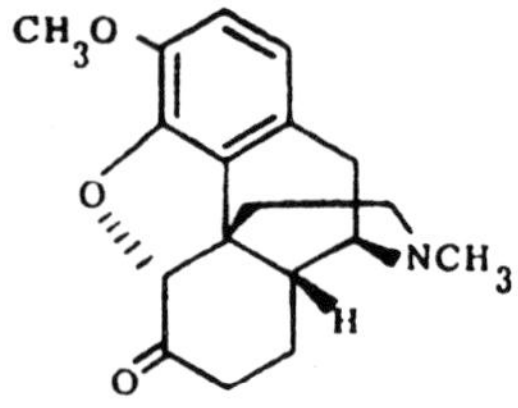

hydrocodone
(dihydrocodeinone)

potent than morphine. Its respiratory depressant action is correspondingly greater, but is less than nauseating and less constipating. The drug may also give rise to tolerance and addiction.[13] There has been extensive study made of the effects of modifying groups on pharmocologic activity as shown in Table 8.2.

The most potent analgesics obtained in this way are methyldihydromorphinone, 14B-acetoxydihydrocodeinone, and N-B-phenylethylnormorphine (Figure 8.6).

THEBAINE

Thebaine also occurs in opium, and in the biosynthesis studies it is converted to codeinone which converts to codeine and finally to morphine. Its concentration range is 0.5-2.5 percent. This alkaloid is currently of great interest because of the need for an alternative source for codeine, whose supply is being threatened. *Papaver bracteatum* Lindl, which produces thebaine in substantial quantities with few other alkaloids, has been studied as a new crop for supplying almost 200 tons of codeine in a year. Structurally it is a methyl enol ether of codeinone[13] (Figure 8.7).

Thebaine contents increase rapidly in roots of young plants of *P. bracteatum* and the concentration in the shoots increase some weeks later. During its second year of growth the highest concentration of thebaine is found in the capsule four to six weeks after flowering. If budding is prevented, the thebaine content remains

TABLE 8.2. Effects of modifying groups on the analgetic activity of morphine

Modification	Increase	Decrease
3-0-Methylation		80-90%
6-0-Methylation	2-fold	
Oxidation of 6-OH	Up to 6-fold	
Reduction of 6, 7-double bond	3-fold	
Quaternization of nitrogen		Complete loss
Nitrogen ring-fission		Complete loss
Opening of 4, 5-oxide bridge		Complete loss
De-N-methyl-N-B-phenylethyl	10-15-fold	
14-Hydroxylation	Substantial	

high in the roots.[13] Alkaloid biosynthesis in *P. bracteatum* is not limited to any part of the plant.[14]

Uses

The alkaloids present in opium in greatest proportion decrease in narcotic properties in the order morphine, codeine, noscopine. Opium and morphine are widely used to relieve pain and are particularly valuable as hypnotics, as, unlike many other hypnotics, they act mainly on the sensory nerve cells of the cerebrum. Codeine is a milder sedative than morphine and is useful for allaying coughing. Both morphine and codeine decrease metabolism and the latter, particularly before the introduction of insulin, was used for the treatment of diabetes. Opium, while closely resembling morphine, exerts its action more slowly and is therefore preferable in many cases (e.g., in the treatment of diarrhea). Opium is also used as a diaphoretic. The habitual use of codeine may in some individuals produce constipation.[1]

FIGURE 8.6. 1-Methyldihydromorphinone, 2-14B-acetoxydehydrocodeinone,3-N-B-phenylethylnormorphine (G. A. Cordell, 1981). Courtesy of John Wiley & Sons, Inc.

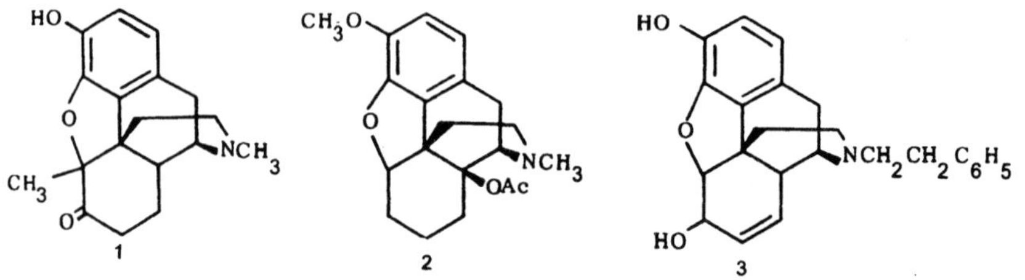

FIGURE 8.7. (G. A. Cordell, 1981). Courtesy of John Wiley & Sons, Inc.

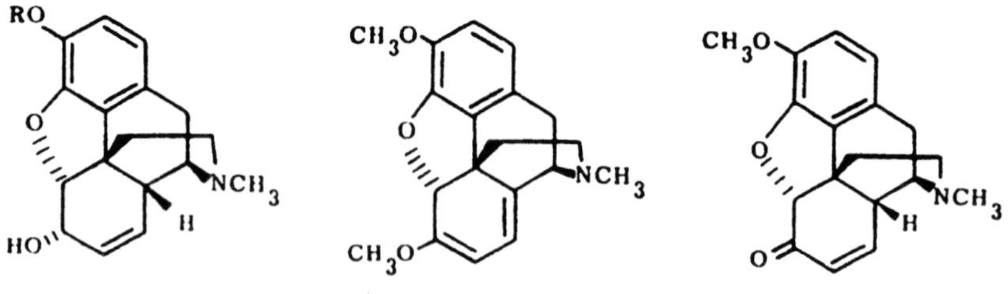

morphine R=H

codeine R=CH_3

thebaine

codeinone

NEOPINE

Neopine, which is isomeric, with codeine gives dihydrocodeine upon catalytic reduction and can differ from codeine in the position of a double bond.[12]

Cordell[13] stated that when 14-bromocodeinone is treated with sodium borohydride, neopine is one of the products together with a new isomer of codeine named indolocodeine and assigned the structure given below. This product is thought to arise by reduction of the intermediate quaternary salt derived from the carbonium ion of 14-bromocodeinone.

Brochmann-Hanssen[11] also observed that the alkaloid neopine is produced when neopinone is subjected to reduction process (Figure 8.8).

Cordell[13] observed that through $^{14}CO_2$ exposure studies and feeding of labeled precursors it was shown that both neopinone and codeinone were precursors of codeine. In addition, these compounds were shown to be minor constituents of the plant. Codeine methyl ether, which was also converted to codeine, was not detected as a natural constituent of the plant (down to 0.02 percent of the thebaine content). Dalton[15] reported that hydrolysis of the enol ether of thebaine yields neopinone, a natural constituent of *Papaver somniferum* L. from which it has been isolated, which subsequently rearranges to codeinone. Reduction of codeinone yields codeine and demethylation of the latter, morphine.

Oripavine, like morphine, is a phenol and on methylation with diazomethane adds CH_2 to give thebaine. Nielsen et al.[16] isolated this alkaloid oripavine in an amount of 0.1 percent for the first time from dried capsules of a variety of opium poppies grown on the

FIGURE 8.8. (G. A.Cordell, 1981). Courtesy John Wiley & Sons, Inc.

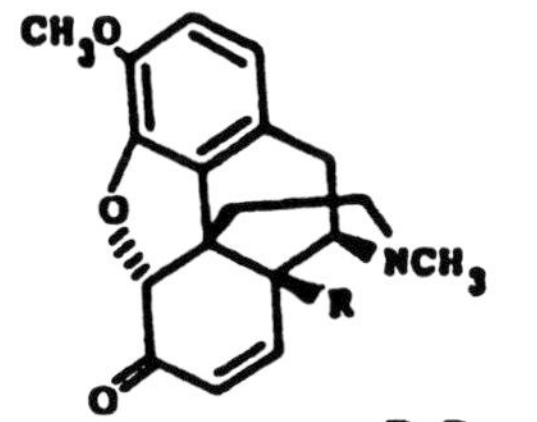

R=Br

14 - bromocodeinone

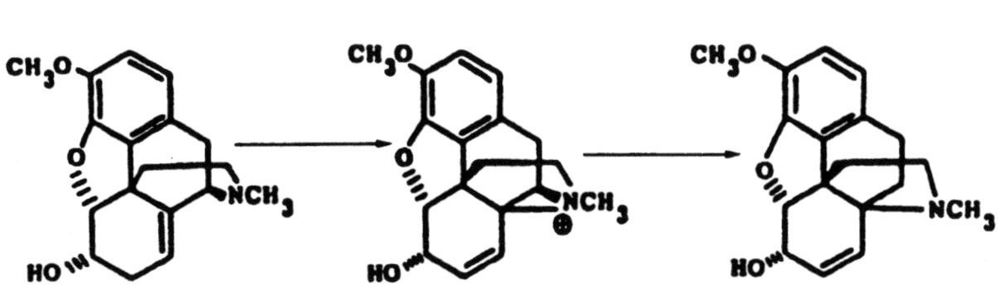

neopine   intermediate quaternary salt   indolinocodeine

island of Tasmania. Oripavine is biosynthesized from reticuline by way of thebaine in *Papaver orientale* and the same pathway is likely to be followed in *P. somniferum* (Figure 8.9).

Oripavine is the main alkaloid in *P. orientale* and *P. bracteatum* hybrids. Oripavidine, a new alkaloid isolated from *Papaver orientale,* was identified as 3.13-didemethylthebaine.[17]

Brochmann-Hanssen[23] has observed that 10-Hydroxycodeine has also been isolated from opium by Small[18] but it is doubted as to whether or not it is a genuine alkaloid[19] (Figure 8.10).

Phillipson et al.[20] isolated the N-oxides of morphine, codeine, and thebaine from *Papaver somniferum.* Each alkaloid forms two N-oxides which have been separated and characterized and also prepared synthetically.

By using ^{14}C-morphine it could be shown that morphine is further degraded to such nonalkaloidal metabolites as thebaol, which is elaborated directly by the plant from thebaine.[21,22]

Brochmann-Hanssen[23] observed that morphinandienone had been isolated from *Croton salutaris* and named salutaridine. Labeled salutardine showed good incorporation into morphine, co-

FIGURE 8.9. (Nielson et al., 1983). Courtesy Planta Medica.

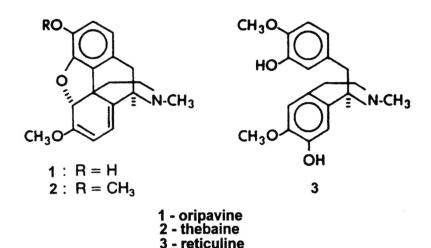

1 : R = H
2 : R = CH₃

3

1 - oripavine
2 - thebaine
3 - reticuline

FIGURE 8.10. (E. Brochmann-Hanssen, 1971). Courtesy Springer-Verlag, Heidelberg.

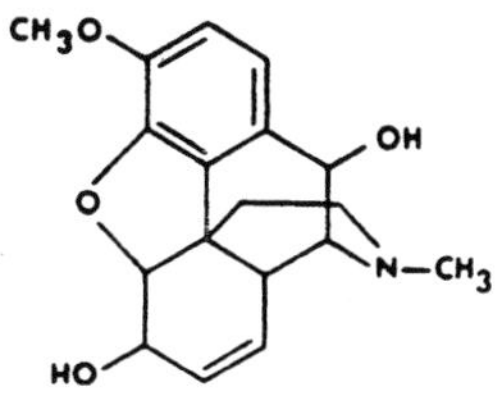

10-Hydroxycodeine

deine, and thebaine and its natural occurrence in opium poppy was demonstrated by an isotope dilution experiment.[24] Oxidative coupling of reticuline to salutardine was carried out in cell-free systems prepared from the opium poppy.[25] Battersby[26] proposed that salutaridine was reduced to a dienol prior to forming the 4,5-epoxybridge. This reduction when carried out *in vitro* with sodium borohydride gave a mixture of two epimers named salutardinol I and salutardinol II. Although both epimers could easily be converted to thebaine *in vitro*, feeding experiments demonstrated that only salutardinol I had the correct stereochemistry for *in vivo* conversion to thebaine by allylic elimination.[27,28]

BENZYLISOQUINOLINE ALKALOIDS

The benzylisoquinoline group of alkaloids in opium poppy includes Papaverine, laudanine, laudanidine, codamine, laudanosine, papaveraldine, and reticuline (Figure 8.11). The pharmacology of some of them is discussed below.

Papaverine occurs in *P. somniferum* L. to the extent of 0.8-1.0 percent and was first obtained by Merk in 1848. The freebase is insoluble in water and only sparingly soluble in ethanol. Cordell observed that papaverine is the most important benzylisoquinoline alkaloid from a pharmacologic point of view. In 1914 Pal demonstrated *in vivo* that papaverine decreases the tonus of the smooth muscle and this had stimulated numerous studies of both a synthetic and pharmacologic nature.

FIGURE 8.11. Benzylisoquinoline alkaloids (G. A. Cordell, 1981). Courtesy of John Wiley & Sons, Inc.

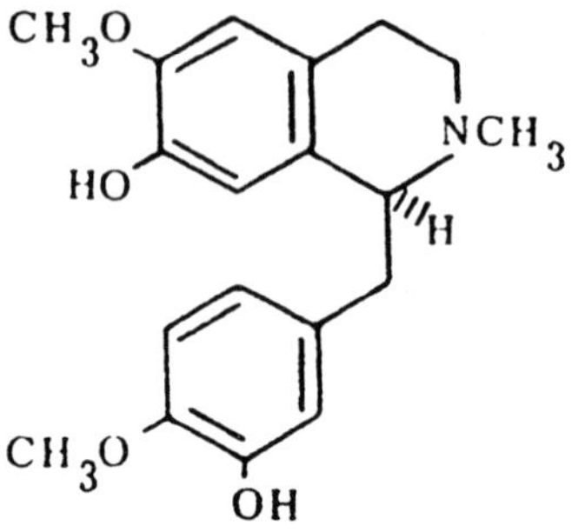

(+)-reticuline

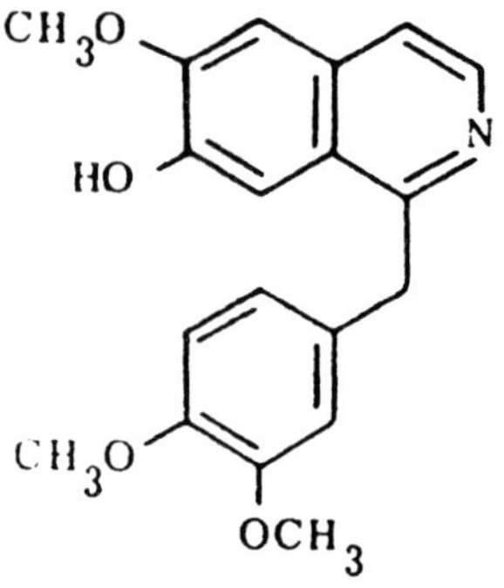

papaverine

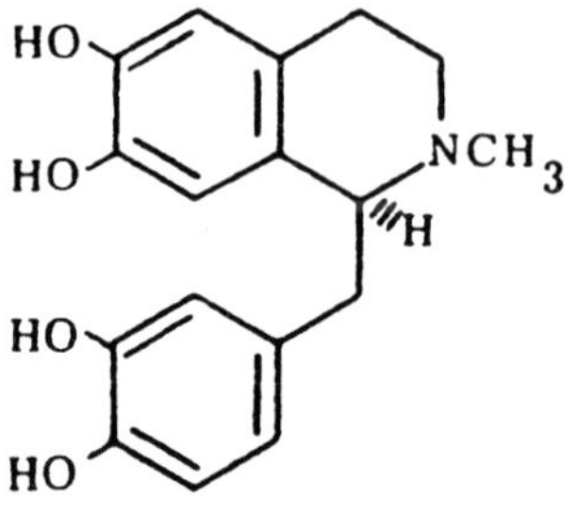

(+)-laudanosoline

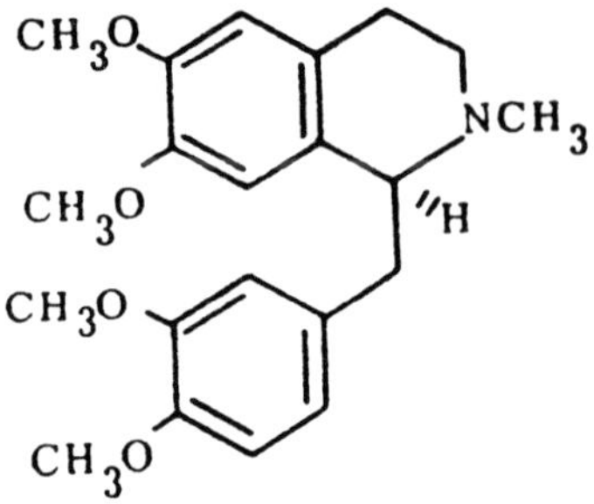

(+)-laudanosine

In humans, papaverine is completely absorbed in the gastrointestinal tract and is metabolized in the liver to 4'-norpapaverine, which is excreted as glucuronate.

Papaverine increases coronary artery flow and causes dilation. In angina pectoris papaverine has a beneficial effect but does not alleviate the pain. The hydrochloride has been used in the treatment of vasospasms accompanying pulmonary embolism and cerebrovascular thrombosis. The glyoxylate salt has spasmolytic and vasodilating properties. Papaverine exerts a strong spasmolytic action on the normal and pregnant uterus. Numerous studies have confirmed the peripheral vasodilating effects of papaverine, and the 3-methyl analogue known as dioxyline, is used as a coronary and peripheral vasodilator. The corresponding methylenedioxy analogue is a smooth muscle relaxant. Sulfonamides potentiate the effects of papaverine, and papaverine increases the inhibition of oxygen uptake by sulfonamides in the liver. Papaverine is neither narcotic nor addictive and side effects include drowsiness and constipation.

LAUDANOSOLINE

Laudanosoline was first isolated in 1871 from opium. Numerous reagents have been used for the 0-demethylation of laudanosine including 48 percent hydrobromic acid and aluminum chloride. Under mild conditions each of the four monophenolic isomers has been obtained. Although considerable information is often obtained from NMR spectra of these alkaloids, a new quaternary aromatic alkaloid may be reduced to tetrahydro species in order to clarify the aromatic region of the spectrum and faciliate determination of the substitution pattern.[13]

LAUDANOSINE

Laudanosine occurs to the extent of less than 0.1 percent in opium and has been synthesized by reduction of papaverine methochloride followed by resolution using quinic acid.[42]

LAUDANINE

Laudanine also occurs in opium and its structure was determined by Spath.[44] When it is methylated with diazomethane it yields (±) laudanosine, so apparently the natural alkaloid is the racemic form, which is unusual.

Laudanosine reduces intraocular pressure in rabbits when administered intravenously. Laudanine had effects similar to strychnine in frogs; in low doses it caused convulsions and larger doses caused paralysis.

RETICULINE

Reticuline was first isolated from *Annona reticulata* (Fam. Annonaceae, order Magnoliales) by Gopinath et al.[36] Labeled reticuline and norreticuline both proved to be very efficient precursors of morphine in the poppy, surpassing norlandansoline in this respect. In 1964 Brochman-Hanssen and Furuya[37] isolated reticuline, which was found to be a normal, minor component of *P. somniferum*. Brochmann-Hanssen and Nielson[38] demonstrated that the reticuline fraction of opium consists of approximately 60 percent of (+) isomer and 40 percent of (−) isomer. Brochmann-Hanssen[11] observed that in the biosynthetic studies of morphine according to speculation a precursor closer to thebaine than norlandanosoline should have an N-methyl group and two 0-methyl groups in the molecule. This new benzylisoquinoline alkaloid, reticuline, fit this requirement. More important, reticuline had the phenolic characteristics and the substitution pattern suitable for intramolecular ortho-para coupling.[24]

Reticuline is regarded as the key intermediate in the biosynthesis of the many alkaloids based on the benzylisoquinoline nucleus and as a result of study of these alkaloids much has been learned of the biosynthesis of reticuline.[13]

Brochmann-Hanssen[11] observed that aphorphines have recently joined the family of opium alkaloids. Nijland[29] reported the isolation of corytuberine and magnoflorine from opium poppy in 1965. Two years later Brochmann-Hanssen et al.[30] isolated isoboldine from opium. These alkaloids have the absolute configuration which relates them to (+)– reticuline and (–)–norlaudansoline (Figure 8.12).

Magnoflorine isolated from *Aristolochia bracteata* seeds decresed arterial blood pressure in rabbits and induced hypothermia in mice. It induced contractions in the isolated guinea pig ileum.[41]

PORPHYROXINE

Merk[31] discovered in 1937 a minor opium alkaloid named Porphyroxine (because of the red color it produced when heated with

FIGURE 8.12. Aphorphines (Brochmann-Hanssen, 1971). Courtesy of Springer-Verlag, Heidelberg.

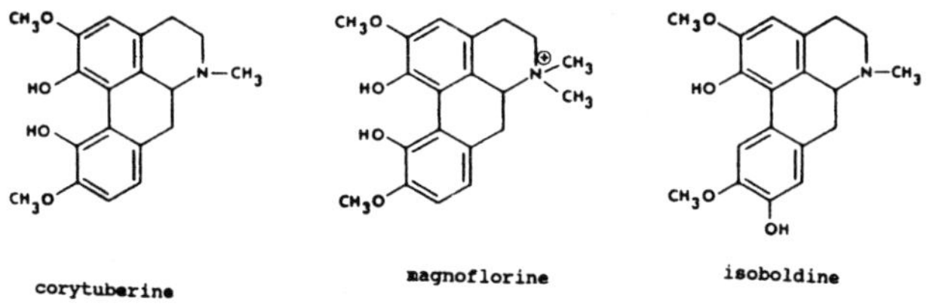

corytuberine   magnoflorine   isoboldine

mineral acids). He suggested that this reaction might be used as a test for opium in forensic cases.[32] The concentration of the alkaloid varies with the source of opium and the color reaction is useful for determining the geographic origin of opium seized in illicit traffic.[33] Pfeifer and Teige[34] determined its empirical formula as $C_{20}H_{21}NO_6$ in 1962. In a series of investigations Pfeifer and Banerjee[35] described six alkaloids which all gave the characteristic red color with mineral acids. They coined the name papaverrubines for these compounds, and their gross structure is illustrated. All papaverrubines are secondary amines. The N-methyl derivatives, several of which are known, do not give the red color reaction with acid (Figure 8.13).

Four such opium alkaloids have been reported, viz. Papaverrubine B, Papaverrubine C, glaudine, and N-methyl-14-0-desmethylepiporphyroxine in addition to Porphyroxine (Papaverrubine D).[11]

THE PROTOPINE GROUP

The alkaloids of this group do not actually contain an isoquinoline nucleus, but they are closely related to those of the berberine type.[42]

Protopine, $C_{21}H_{19}NO_5$, was first isolated from opium in 1871 and is said to have been found in all species of Papaveraceae. In *P. somniferum* it is present in minute amounts while in others, e.g., *Dicentra spectabilis* Lem, it is the chief alkaloid.

Cryptopine, $C_{21}H_{23}ON_5$, was first isolated in 1867 from opium which was the only known source when Perkin[43] completed his classical research on its structure in 1916. It has since been found in

FIGURE 8.13. Gross structure of Papaverrubines. (Brochmann-Hanssen, 1971). Courtesy of Springer-Verlag, Heidelberg.

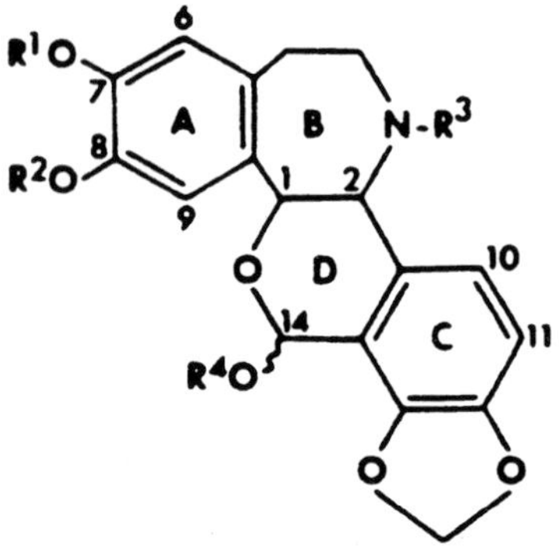

Papaverrubines

other plants, e.g., species of *Corydalis* and *Dicentra.* Cryptopine, m.p. 221°, is sparingly soluble in most organic solvents except chloroform. The protopines have thus far been of little interest pharmacologically.

Protopine-type alkaloids slow the heartbeat and have antifibrillatory properties. They also increase the coronary artery flow. Protopine and cryptopine stimulate the uterus but the effect is of brief duration.[40]

PHTHALIDEISOQUINOLINE ALKALOIDS

As stated by Cordell,[13] there are about 15 well-characterized phthalideisoquinoline alkaloids known based on the skeleton isolated mainly from Papaveraceae and Berberidaceae.

Narcotine is one of the major alkaloids of opium *Papaver somniferum* and was originally isolated by Derosne in 1803. The gross structure of narcotine was established by hydrolysis with dilute sulfuric acid to cotarnine and opiaic acid and by zinc-acid reduction to hydrocotarnine and meconine (Figure 8.14).

FIGURE 8.14. Narcotine (G. A. Cordell, 1981). Courtesy of John Wiley & Sons, Inc.

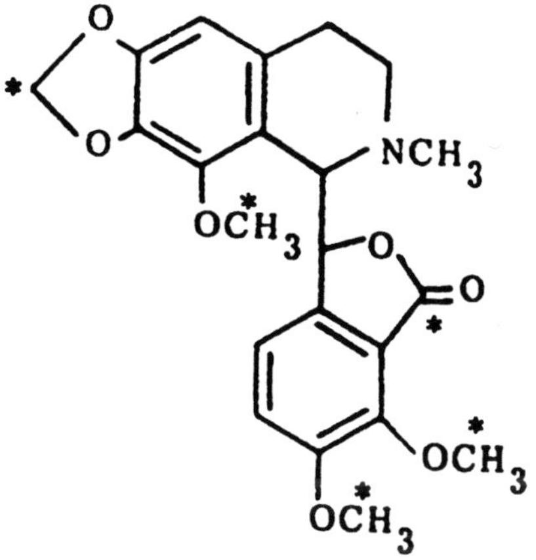

narcotine

PHARMACOLOGY

Narcotine is a mild antitussive and has a relaxant effect on smooth muscles. β-narcotine N-oxide is a more effective antitussive than dihydrocodeine. Narcotine potentiates the mitotic effects of colchicine, although it has no action alone.[13]

Belying its name, the alkaloid has only a mild narcotic action and is a much weaker analgesic than morphine and codeine.[40]

NARCEINONE

Chaudhury and Thakur[39] isolated a new alkaloid named Narceinone from a new strain of *P. somniferum* developed at CIMAP (Lucknow) along with other known alkaloids including narceine. The authors claim that narceinone, a new secophlhalideisoquinoline alkaloid, has been identified in the unlanced dried capsules of *Papaver somniferum* as 14-oxonarceine from spectral analysis and by the biomimetic oxidation of narceine. Narceinone is the fourth example of the occurrence of diketosecophthalideisoquinoline alkaloid in nature (Figure 8.15).

FIGURE 8.15. Narceinone, an alkaloid from *P. somniferum*. Reprinted with permission of Phytochemistry Vol. 28(7). Copyright 1989. Pergamon Press, plc.

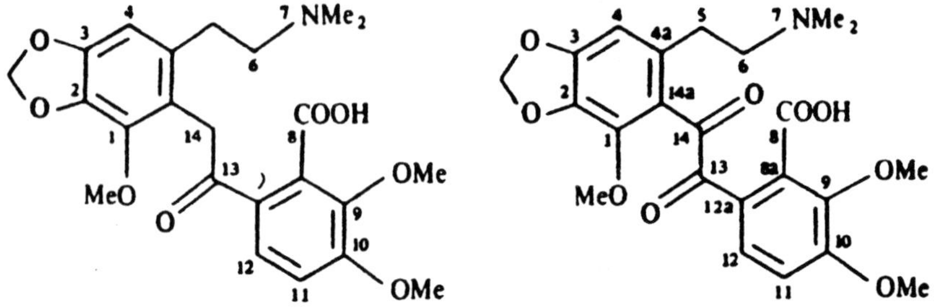

narceine

narceinone

This is however not the conclusive story of chemistry of opium alkaloids. Many workers are actively engaged in this project and their investigations may reveal some more interesting unreported alkaloids, intermediates, or pathways. The work on enzymes has recently been initiated and this may throw light on the identification of specific enzymes responsible for biosynthesis of different alkaloids in opium poppy.

REFERENCES

1. Trease, G. E. and Evans, W. G. Pharmacognosy, 12th ed. Bailliere Tindal, London, 1983.

2. Fairbairn, J. W. and Williamson, E. W. Meconic Acid as a Chemotaxonomic Marker in the Papaveracea,. Phytochemistry, 17, 1978, pp. 2087-2089.

3. Fairbairn, J. W. and Steele, M. J. Meconic Acid and Alkaloids in *Papaver somniferum* and *P. bracteatum*, Planta Medica, 41, 1981, pp. 55-60.

4. Henry, T. A. The Plant Alkaloids. 4th edition. The Blakiston Co., Philadelphia, 1949.

5. Gulland, J. M. and Robinson, R. J. Morphine Group I. Discussion of the Constitionalar Formula. Chem. Soc. 123, 980, 1923.

6. Gulland, J. M. and Robinson, R. Constitution of Codein and Thebaine, Mem. Proc. Manchester, Lit. Philos. Soc., 69, 1925, pp. 79-86.

7. Thakur, R. S. 'Opium: The Chemistry and Uses,' in 'The Opium Poppy,' ed. Husain, A. and Sharma, J. R. Central Institute of Medicinal and Aromatic plants, Lucknow (India), 1983.

8. Taylor, F. O. Allen's Commercial Organic Analysis. Vol. VII. (ed. C. A. Mitchell). J and A Churchill Ltd., London, 1929, pp. 655-751.

9. Santavy, F. Papaveraceae alkaloids in 'Alkaloids' by R. H. F. Manske, Vol. XII. Academic Press, New York, 1970.

10. Santavy, F. Papaveraceae Alkaloids, in 'Alkaloids,' edited by R. H. F. Manske and R. Rodrigo, Vol. XVII. Academic Press, New York, 1979 .

11. Brochmann-Hanssen, E. Aspects of Chemistry and Biosynthesis of Opium Alkaloids, in 'Pharmacognosy and Phytochemistry' First International Congress, Munich, 1970, edited by H. Wagner and L. Horhammer. Springer-Verlag, Berlin, Heidelberg, New York, 1971.

12. Bentley, K. W. The Morphine Alkaloids, in 'Chemistry of Alkaloids' (ed. S. W. Pelletier). Van Nostrand Reinhold, New York, 1970.

13. Cordell, G. A. Introduction to Alkaloids. A biogenetic Approach. John Wiley and Sons, Inc., New York, 1981.

14. Mothes, K., Schutte, H. R. and Luckner, M. Biochemistry of Alkaloids. Deutscher Verlag der Wissenchaften, Berlin (DDR), Germany, 1985.

15. Dalton, D. R. The Alkaloids. Marcel Dekker Inc., New York, 1979.

16. Nielson, B., Roe, J. and Brochmann-Hanssen, E. Oripavine–A New Opium Alkaloid, Planta Medica, 48, 1983, pp. 205-206.

17. Israilov, I. A., Denisenko, O. N., Yunusov, M. S., Yunusov, S. Yu and Muraveva, A. D. Khim. Prir. Soedin, 714 (1977); CA 88, 170357f, 1978.

18. Small, L. F., cited by B. Witkop and S. Goodwin. J. Am. Chem. Soc., 75, 1953, pp. 3371.

19. Kuhn, L. and Pfeifer, S. Pharmazie, 18, 1963, p. 819.

20. Phillipson, J. D., Handa, S. S. and El-Dabbas, S. W. N-oxides of Morphine, Codein and Thebaine and Their Occurence in Papaver Species, Phytochemistry, 15, 1976, pp. 1297-1301.

21. Kunitomo, J., Yuge, E., Nagai, Y. and Fujitani. Chem. Pharm. Bull., 16, 1968, p. 4066.

22. Reisch, J., Gombos, M. Szendrei, K. and Novak, I. Arch. Pharm. (Weinheim, Ger.), 307, 1974, p. 814.

23. Brochmann-Hanssen, E. 'Biosynthesis of Morphinan Alkaloids,' in The Chemistry and Biology of Isoquinoline Alkaloids, ed. Phillipson et al. Springer-Verlag, Berlin, Heidelberg, 1985.

24. Barton, D. H. R. and Cohen, T. Some Biogenetic Aspects of Phenol Oxidation. Festschr stoll. Birkhauser Basel, 1957, pp. 117-143.

25. Hodges, C. C. and Rapoport, H. Enzymic Conversion of Reticuline to Salutaridine by Cell Free Systems. Biochemistry, 21, 1982, pp. 3729-3734,.

26. Battersby, A. R. Tilden Lecture. The Biosynthesis of Alkaloids. Proc. Chem. Soc., 1963, pp. 189-200.

27. Barton, D. H. R., Kirby, G. W., Steglich, W., Thomas, G. M., Battersby, A. R., Dobson, T. A. and Ramuz, H. Investigation on the Biosynthesis of Morphine Alkaloids, J. Chem. Soc., 2423, 1965, p. 2438.

28. Barton, D. H. R., Bhakuni, D. S., James, R. and Kirby, G. W. Phenol Oxidation and Biosynthesis Part I. XII Stereochemical Studies Related to the Biosynthesis of the Morphine Alkaloids, J. Chem. Soc. (C), 1967, pp. 128-132.

29. Nijland, M. M. Pharm. Weekbl. 100, 1965, p. 88.

30. Brochmann-Hanssen, E., Nielsen, B. and Hirai, K. J. Pharm. Sci., 56, 1967, 754.

31. Merk, E. Ann. Pharm., 21, 1937, p. 201.

32. Merk, E., in 'Jahresber. uber Forlschritte der chemie und Mineralogie,' Berzelius J., Vol. 24, Tubingen, Germany, 1845, p. 399, Bulletin on Narcotics, 4, 1952, p. 15.

33. United Nations Secretariat, United Nations Document, ST/SOA/Ser. K/65.

34. Pfeifer, S. and Teige, J. Pharmazie 17, 1962, p. 692.

35. Pfeifer, S. and Banerjee, S. K. Pharmazie 19, 1964, p. 286, Arch. Pharm. 298, 1965, p. 385.

36. Gopinath, K. W., Govindachari, T. R., Pai, B. R. and Viswanathan, N. The Structure of Reticuline, a New Alkaloid from Annona Reticulat, Chem. Ber., 92, 1959, pp. 776-779.

37. Brochmann-Hanssen, E. and Furuya, T. A New Opium Alkaloid. Isolation and Characterization of (+)-1-(3'-hydroxy-4'-methoxybenzyl)-2 methyl-6methoxy-7-hydroxy-1,2,3,4-tetrahydro-isoquinoline [(+) reticuline], Planta Med., 12, 1964, p. 328.

38. Brochmann-Hanssen, E. and Nielsen, B. (+)-Reticuline–a New Opium Alkaloid. Tetrahedron Letters. No. 18, 1965, pp. 1271-1274.

39. Chaudhury, P. K. and Thakur, R. S. Narceinone, an Alkaloid from *Papaver somniferum*. Phytochemistry. Vol. 28, No. 7, 1989, pp. 2002-2003.

40. Shamma, Maurice. 'The Isoquinoline Alkaloids.' Academic Press, New York, 1972.

41. Kamal, E. H. EL TAHIR–Pharmocological Actions of Magnoflorine and Aristolochic Acid-1 Isolated from the Seeds of Aristolochia Bracteata. Inter. J, of Pharmacognosy, 29, (2), 1991, pp. 101-110.

42. Swan, G. A. An Introduction to Alkaloids. John Wiley and Sons, Inc., New York, 1967, pp. 112-113.

43. Perkin, W. H., Jr., J. Chem. Soc. London, 109, 1916 p. 815; 115, 1919, p. 713.

44. Spath, E. Constitution of laudanine Monatsch 41, 297, 1920.

Chapter 9

Biosynthesis and Physiology of Opium Alkaloids

It has been a curiosity to know how the tiny little poppy seed, which is free from alkaloids, could be a fountainhead for more than 40 alkaloids in the *Papaver somniferum* which grows out of it. Are these alkaloids formed in the root or leaf or latex of the seedling or in the mature plant? Which one is the first and how is it converted into the others? What are their pathways? How and why are so many of these alkaloids accumulated in *P. somniferum* or other species of *Papaver* and what is the role of these alkaloids in the plant and outside the plant?

These are the challenging queries enagaging the attention of many workers. The study of the molecular structures of alkaloids and of the substances occurring along with them in the plant throws some light on the process by which the opium alkaloids are synthesized in plants. The use of radioactive techniques and isolation of various enzymes from the poppy plant have also helped to give a broad outline of biogenetic process.

Robinson[106] was a pioneer researcher in fundamental chemistry and has put forward many important ideas on alkaloid biogenesis. The speculations were based largely upon structural similarities within the alkaloid series and also upon the relations of alkaloids to simpler natural products. Although such approaches were not claimed to prove that plants necessarily followed the suggested biosynthetic schemes, nevertheless, the proposals have given help to structural studies on new alkaloids and led to experiments on living plants.[114]

The opium alkaloids are synthesized from amino acids, phenylalanine, and tyrosine. The amino acids are the product of protein degradation in germinating seeds or they are synthesized in the leaves, roots, etc., from organic acids.[1] The findings of numerous research workers engaged in this field answer some of the questions of the inquisitive mind.

SITE OF FORMATION OF ALKALOIDS

The site of formation of alkaloids in plants has been the subject of interest for quite some time. The histochemical approach of some workers revealed that alkaloids nearly always occur in living cells.

Using physiological methods some workers have shown that in Solanaceous plants alkaloids are produced in the germinating roots and translocated to other parts of the plant. The presence of alkaloids in seedlings and roots of *Atropa belladonna* and *Datura stramonium* has been demonstrated but interestingly their seeds also contain the alkaloids.[2] The presence of alkaloids in the germinating radicle of *Physochloaina preleata* (D) Miers, a solanaceous plant whose seeds are free from alkaloids like that of *Papaver somniferum* L., has also been reported.[3]

In *Papaver* species, due to latex oozing out of the cut portions, grafting is not possible to determine precisely the site of formation of alkaloids. The only evidence available on this aspect is from biosynthetic experiments performed on intact plants, germinating seeds, seedlings, isolated organs, or tissue culture.[4] Most of the plants of *Papaver* contain hydrophenanthrene, phthalideisoquinoline, or benzylisoquinoline group of alkaloids.[5]

The earliest recorded investigation on the localization and distribution of alkaloids in opium poppy are those of Clautriau[6] and Errera et al.[7] According to them the young plant is not poisonous until it has attained the height of 10-15 cm when it contains appreciable traces of morphine which then occurs in the whitish latex but not at shoot or root apex. Clautriau[6] also reported that the seeds are devoid of alkaloids. Kerbosch,[8] in a detailed investigation, confirmed that the seeds of opium poppy are free of alkaloids but during germination narcotine is the first alkaloid to appear. According to him, codeine, morphine, and papaverine may appear when seedlings are 5-7 cms high. He was of the opinion that the sprouting seeds manufactured narcotine from their protein reserve even when germinated in a nitrogen-free environment. Codeine developed next. Valaer[9] observed that alkaloids occur in seedlings pulled up three weeks after planting seeds. These tiny seedlings were about 2-5 cms tall and proved to contain morphine.

Mika[10] confirmed the findings of Kerbosch, of absence of morphine or other alkaloids in the seeds of opium poppy: 5g of air-dry seeds were analyzed and no alkaloid was detected.

Opium alkaloids have not been reported to occur in any tissue other than laticifers. Where there are no laticifers in any particular organ of opium poppy no alkaloids may be found.[11] Kapoor[12] reported that stamens and seed are devoid of laticiferous vessels and at the same time are devoid of alkaloids. The young seedlings of opium poppy collected before the cotyledonous stage contain neither laticiferous vessels nor alkaloids. The young seedling whose cotyledons have opened possessed laticiferous vessels but gave no reaction for alkaloids. The cell contents at this stage gave precipation with Iodine-potassium iodide (I_2KI) but no specific reaction for morphine could be observed. This precipation with I_2KI may be due to protein which also gives a similar reaction to alkaloids. Fairbairn and Kapoor[11] observed that the first appearance of green tissue in the seedlings coincides with the appearance of laticifers; the later arise a short time before alkaloids can be detected. Laticifers appear to be associated with photosynthetic tissue and make the first appearance in the seedling when the first green cotyledons appear.

Massicot[13] studied the alkaloid contents during eight days of germination and concluded that alkaloids are synthesized by the whole plant and not only by the root.

Fairbairn and Kapoor[11] observed that alkaloids of *P. somniferum* are restricted to the laticifers. They showed that those parts of the plant in which no laticifers occurred gave no reaction for alkaloids. Further, laticiferous tissue is particularly abundant in the outer mesocarp of the capsule two weeks after the fall of the petals; and it is from this region and at this stage of development that the maximum yield of latex is obtained. The restriction of the alkaloid to highly specialized laticiferous vessels is in contrast to the situation in many other alkaloid-containing plants such as *Atropa, Nicotiana,* and *Datura* species. In these plants, the alkaloids occur in "normal" living cells and especially those where metabolic activity is high, such as shoot and root apices.[2] It can be assumed that the tissues of such plants are immune to any toxic effects of the alkaloids.[17] The situation in the poppy plant suggests, on the other hand, that normal tissues are not able to adjust themselves to the presence of alkaloids, at least in detectable quantities. In their opinion, if this sensitivity to alkaloids does occur, then it appears to be most marked in one critical phase of the life history, namely, the formation and development of the seed, as it has been shown that both laticifers and alkaloids are absent from the stamens, ovules, seed, and the young seedlings. Meiotic and mitotic activity is high during this phase and morphine has been shown to possess antimitotic activity similar to colchicine.[18] This suggestion that the alkaloids are completely absent during this stage of the life history because they have toxic mitotic effects may however appear too teleological. An alternative suggestion is that the absence of alkaloids is related to the absence of photosynthetic tissue. Chlorenchyma is absent from the stamens, ovule, and the seed; in the seedling the appearance of green tissue coincides with that of laticifers and of detectable quantities of alkaloids. Furthermore, in the capsule wall the laticifers are most abundant in the mesocarp region which is rich in chloroplast and narrower in the regions of placenta, remote from the chlorenchyma. In week 2 after the fall of petals the green capsule reaches its maximum surface area and presumably photosynthetic activity is at its maximum: this peak corresponds to the time of maximum yield of latex and alkaloids.

It may be that alkaloids are necessary by-products of metabolic activity, especially that associated with photosynthesis, but because of their toxic effects in normal cells they are immediately secreted into laticifers. If this is so then the site of synthesis would be the leaf and other photosynthetic tissues rather than in the root as occurs in certain members of the Solanaceae.

This in turn may account for the fact that the laticifers always occur in the phloem and never in immediate contact with xylem vessels. Translocation would therefore be "downward" from photosynthetic tissue through the phloem rather than upward from the root, through the xylem as occurs in Nicotiana.[19]

Crane and Fairbairn[20] observed that the seedlings of commercial "blue" strain of *Papaver somniferum* when germinated showed by the third day traces of morphine, codeine, thebaine, papaverine, and narcotine. On the fifth day there were traces of reticuline and on the sixth day traces of narcotoline were detected.

Although there is no anatomical connection between the laticiferous vessels and the seed, the presence of radioactive derivates of morphine was detected in seed by Fairbairn and Paterson.[60] They fed ^{14}C morphine to the growing poppy plants and collected seed a few days after and detected radioactivity in them (Table 9.1).

Methanol extracts of seeds from plants fed with carbon[14]-labeled tyrosine also showed similar radioactive ninhydrin spots which contained about 75 percent of the total activity of the extracts.

MORPHINE AND SEED DEVELOPMENT

In persuance of this finding Fairbairn and El-Masry[28] continued deeper study of the seed and its development. Vigorous acid hydrolysis or pepsin digestion of ground poppy seed led to the production of alkaloid-like substances including codeine.[28] Fermentation of ground seeds or germination of whole seeds for one to two days also led to the production of similar compounds as well as papaverine. Radioactive morphine was fed to latex of the capsules on plants and the seeds collected later were germinated in similar conditions

TABLE 9.1. Radioactivity in menthol extracts of poppy seeds collected after feeding the plants with "C-morphine"

[14]C-morphine fed	Seeds collected after feeding (days)	Total radio-activity in methanol extract of seeds (d.p.s.)	Radio-activity incorporated (percent)
5 mg containing 680 d.p.s.	3	20	3.0
23 mg containing 2,040 d.p.s.	5	116	5.7

and the alkaloid-like substances shown to be radioactive. Further-more, the ovules were shown to contain radioactive alkaloid-like substances (without degradative treatment) as well as radioactive non-alkaloidal compounds.

It has been suggested by the authors that morphine is rapidly metabolized in the latex into a series of compounds, some of which are alkaloid-like and others non-alkaloidal, ethanol-insoluble "bound" forms. The bound forms are stored in the seeds and bro-ken down into smaller alkaloid-like substances during germination. The authors are of the opinion that poppy seeds contain molecules (presumably larger than normal alkaloids) which on degradative treatment produce alkaloid-like substances, one of which is co-deine. Even after feeding radioactive morphine to the latex, radioac-tive alkaloid-like substances and ethanol insoluble ("bound") forms) appeared in ovules and seeds. Germination of the mature seeds led to the breakdown of the bound forms to produce radioac-tive alkaloid-like substances. Morphine is rapidly metabolized in the latex[35] into a series of compounds, some of which are alkaloid-like and ethanol soluble and others non-alkaloidal, ethanol insolu-ble, and probably fairly large molecules. Some of these are trans-ported to the developing seeds and stored in them mainly as the large molecules. During germination these compounds breakdown to form the smaller alkaloid-like substances. It is possible that some other alkaloids from the latex are also metabolized similarly since codeine and papaverine are detectable in germinated seeds before *denovo* synthesis is likely.

These results add to the evidence that morphine is not an end-product substance but is metabolized into a series of new com-pounds.[28]

To understand the functions of "bound" forms of alkaloids some capsules were regularly incised during their development so as to remove as much latex as possible.[29] As such, the seeds within these capsules were deprived of a significant amount of latex products during their development. The ripe seeds did not look different from those of a control group of plants but showed significant differences in viability. This is shown in Table 9.2. Only about half of the "bled" seeds produced roots and these emerged slowly and were often deficient in root hairs. In those seedlings which pro-

TABLE 9.2. Viabilities of seeds and chlorophyll content of seedlings from "bled" capsules and from controls

Days after ger- mina- tion	Emergent roots (%)		Roots possessing hairs (%)		Emergent cotyledons (%)		Fully opened cotyledons (%)	
	Contr.	Bled.	Contr.	Bled.	Contr.	Bled	Contr.	Bled
1	75	1	–	–	–	–	–	–
2	93	37	92	4	–	–	–	–
3	97	48	97	36	83	15	52	7
5	98	50	98	36	98	36	83	20
7	98	50	98	36	98	48	97	48

Chlorophyll content µg/100 seedlings.

Days after germination	Chlorophyll a		Chlorophyll b	
	Normal	Bled	Normal	Bled
5	57.5	33.5	31.6	21.2
8	83.5	64.7	35.0	24.1
10	97.7	97.6	52.7	47.0

Reprinted with permission from *Phytochemistry, Vol. 7*, pp. 181-187. (Fairbairn & El-Masry) © 1968, Pergamon Press.

duced cotyledons the chlorophyll content was also deficient. However, the seedlings which survived and produced their own leaves developed into normal healthy plants. It is suggested that the "bound" forms of the alkaloids are essential to the early stages of seed development. Later, when the leaves open out, the plant begins to produce its own alkaloid and can then develop normally.

Wold[96] also found that polysaccharide fraction of the capsule contained bound morphine and codeine. The alkaloids appeared to be bound to the polymer by two different types of linkage. He demonstrated that alkaloids are bound also to the H_2O-soluble high MW fraction isolated from poppy capsules freed from seeds.

LATEX AND ALKALOID VESICLES

Radioactive 3,4-dehydroxyphenylalanine has been found to be much more efficiently incorporated into the alkaloid than is tyrosine or glucose in poppy latex.[26] The authors isolated radioactive morphine and four other alkaloids whose purity was established by radioactive measurements. The fact that such sophisticated synthesis takes place in isolated latex indicates that the latter is cytoplasmic with adequate organelle activity. If such a synthesis involving many enzyme systems does take place in the latex, it may well be carried out at particular sites. The necessary organelles should therefore be present in the isolated latex from the unripe capsules.

In the persuance of this investigation the isolated latex was mixed with mannitol buffer and pellets were collected after spinning for 10 minutes at 1,000g, 3,000g and 11,000g.[26]

Fairbairn and Djote[21] reported that the pellet produced by centrifuging suitably diluted latex at 1,000-3,000g could synthesize morphine from dihydroxyphenylalanine *in vitro* and preliminary studies indicated that a lutoid-like organelle is involved in this synthesis. Further metabolism of morphine probably occurs in the stem latex and its 3000g fraction rather than in the capsule latex. It has been reported by workers in Mothes's laboratory that biosynthesis of alkaloids from tyrosin was observed in isolated leaf, capsule, and the latex of *P. somniferum*.[22] All these parts were able to synthesize alkaloid but latex was considered the best.[23]

SUBCELLULAR LOCALIZATION OF ALKALOIDS

Fairbairn et al.[24,25,26] showed that the pellet obtained by centrifuging the latex was capable of performing certain biosynthetic reactions leading to morphine. The pellet consists of vacuolar particles, and the name proposed for these particles was "alkaloid vesicles." These vesicles were suggested to have the functions of (1) storage of alkaloids, (2) biosynthesis and catabolism of alkaloids, and (3) translocation of alkaloids. After centrifuging stem and capsule latex at 1000g, 95-99 percent of the alkaloids were found in the pellet which consisted mainly of "alkaloidal vesicles."[25] The alkaloids appear to be stored in the vacuolar sap of the vesicles rather than membrane bound, and in this respect the vesicles behave as normal vacuoles. It is indeed a remarkable finding that a specific organelle in the plant has been credited with these three functions.[4] The stem latex and vesicles are translocated into the capsule during its rapid expansion after petal fall. During this time the morphine itself is being synthesized and metabolized in the vesicles (more rapidly in the stem vesicles than in those of capsule) and the metabolites pass out of the latex into the pericarp, with a significant amount appearing in the ovules. The vesicles are therefore not merely passive accumulators of alkaloids.[25]

As stated earlier, the latex contains viable mitochondria and a heavier fraction contains larger organelles. When the latex is centrifuged at about 3,000g these heavy particles form a pellet at the bottom of the tube; the supernatant fluid contains the mitochondria, ribosomes, enzymes, and other small molecules, but surprisingly, no alkaloids are present. These are always associated with the heavy organelle fraction provided these are not plasmolyzed.[29] Interestingly, the heavy fraction when separated from the rest of the latex, will synthesize morphine from radioactive dihydroxyphenylalanine. This strongly suggests that the entire synthesis of the opium alkaloids from simple amino acids is carried out by specially designed organelles in this plant. The author suggests that the bound forms which are produced from the morphine are later released from the particles for translocation to the seeds and other parts of the plant.

The polysaccharide fraction of the pericarp and seed of *P. somniferum* contained "bound" forms of morphine which were derived from radioactive morphine fed to the living plants. Bound forms of codeine, thebaine, and some unidentified alkaloid-like compounds were detected in the pericarp and "bound" thebaine occurred in the pericarp of *Papaver bracteatum* also. The complexity and molecular weight of the bound alkaloids seemed to increase during ripening and it is suggested that these substances represent transitional forms in the metabolism and translocation of morphine from latex to seed.[45]

Dickenson and Fairbairn[27] observed that the latex of *P. somniferum* contains abundant small vesicles. Their ultrastructure was studied in tissue sections from adult plants and in sections of sequential fractions of centrifuged latex. The vesicles were found to exist in two forms, the first with a smooth but progressively granulated outer membrane and the second, probably derived from the first, with adherent cap-like structures which in the heavier centrifuged fraction possessed a zonally ordered interior.

The vesicle fractions were found active in synthesizing morphine. *Papaver somniferum* also contains an organelle which was found to resemble a complex organelle present in the latex of *Hevea brasiliensis*. Its function is not yet known.

The subcellular distribution of MLPs (major latex proteins) was examined in developing laticifers of poppy using ultra structural immunolocalization. MLP is a group of low-molecular-weight peptides. In mature laticifers the MLPs are localized in a subpopulation of the latex. In some laticifers the protein appears to be compartmentalized within individual vesicles. Associated with the mature laticifers are cells which also contain MLPs in a large central, apparently protein-filled vacuole. These cells are interpreted to be developing laticifers.[116]

From the distribution of the two forms of alkaloidal vesicles it is possible that the capped form may represent the persistent mature form of its precurser, the transitory juvenile, smooth, or granulated vesicle. It is speculated that the capped form showing structural zonation is the dominant active site for morphine biosynthetic activity.

DIURNAL VARIATIONS OF ALKALOIDS

Waller and Nowacki[4] observed that determining the fluctuations that occur during a single day indicates active metabolization of alkaloids in a plant. They call it catabolism of alkaloids with metabolites unknown.

Pfeifer and Heydenreich[30] analyzed poppy plants collected at six-hour intervals and showed that marked variations in the contents of all the alkaloids occurred. Fairbairn and Wassel[94] observed marked variation in the contents of all three alkaloids, thebaine, codeine, and morphine, at intervals of one to three hours, suggesting that these alkaloids play an active part in metabolism rather than occurring as a slowly accumulating amount of waste matter. Secondly, though morphine has been shown to occur as the irreversible end-product of the sequence thebaine-codeine-morphine, it has been found to decrease markedly at certain times during the day. This suggests that it is periodically converted into a non-alkaloidal molecule which may remain in the latex or pass out into other tissues of the capsule.

Waller and Nowacki[4] have observed that morphine, which is mostly found in the roots of young poppy plants, gradually increases in concentration in leaves. At the stage of fruit formation morphine disappears from leaves and accumulates in high levels in the fruit capsule. In their opinion, diurnal variation of alkaloid concentration in the plant shows complementarity between thebaine and codeine, but both drop just before the morning rise in morphine (Figure 9.1). This is consistent with the known biosynthetic pathway: thebaine-codeine-morphine. Morphine is always prodominant and more morphine is made than can be accounted for by simple conversion of thebaine and codeine, both of which disappeared. A reasonable hypothesis is rapid biosynthesis of the thebaine and conversion to codeine and codeine to morphine followed by rapid conversion of morphine to non-radioactive substances. Such a process would result in the observed daily concentrations of morphine.

The rapid turnover of the alkaloids and periodic disappearance of morphine was confirmed[32] by feeding L-Tyrosine $U^{-14}C$ in solution to the phloem region of the pedicels of poppy plants and analysis of latex from the corresponding capsule. Incorporation of tyrosine into the alkaloids was at least 4 percent and in view of the rapid turnover, probably considerably exceeded this figure. The synthesis of these alkaloids from tyrosine cannot therefore be considered aberrant but the results also indicate that there may be an additional precurser for alkaloids.

It was also observed that after ten minutes a significant incorporation of radioactivity into alkaloid had taken place, and the isoquinoline alkaloid had the highest activity. Morphine disappears from the latex instead of accumulating and a fall in the radioactivity of morphine, codeine, and thebaine is sometimes accompanied by an increase in the amount of alkaloids present. Since isolated latex was used, possibly morphine is transformed into non-alkaloidal molecules in the latex and there may be an additional non-tyrosine route to the alkaloids.[31] Earlier Kleinschmidt and Mothes[22,23] fed radioactive tyrosine and radioactive glucose to isolated poppy latex and showed that radioactive alkaloids could be recovered 24 hours later.

Fairbairn and El-Masry[35] fed morphine U-14C to the phloem

FIGURE 9.1. Diurnal changes in contents of alkaloids of whole *P. somniferum* plants. Reprinted with permission from *Phytochemistry*, 3 (Fairbairn & Wassel). ©1964. Pergamon Press and Plenum Publishing Corporation, New York.

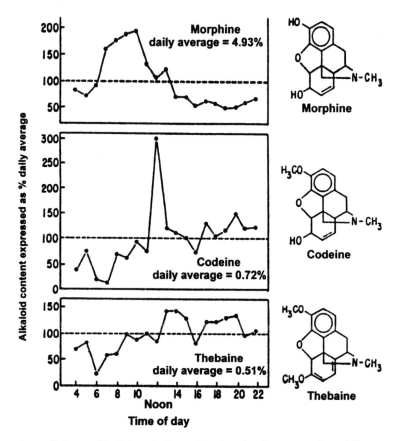

region of the pedicel just below the developing capsule of *P. somniferum* L. Examination of the capsule latex at intervals after feeding showed that there had been rapid absorption of the morphine into the latex which was in marked contrast to the rate for tyrosine. The morphine was rapidly metabolized in the latex to form two non-alkaloidal polar substances and the bulk of these were rapidly translocated out of the latex. They and possibly other related derivatives appeared in the pericarp and ovules and at later stages seemed to form methanol-insoluble substances or were metabolized as $^{14}CO_2$.

Some of the fed morphine was localized at the site of injection for a time and was transformed into other Dragendorff positive substances, amino acids, and sugars. These may be break-down products produced by extra-laticiferous enzymes and therefore do not represent the normal metabolism of morphine.

While studying the fate of [^{14}C]-labeled morphine in *P. somniferum* it was found that part of radioactivity could be detected among sugars and amino acids. None of these radioactive metabolities were identified. However, extracts of *P. somniferum* showed the presence of several polar unidentified Dragendorff positive spots. The high rate of morphine turnover led the workers to conclude that alkaloid has an active role in metabolism: one of its roles is to serve as a methylating agent.

Fairbairn et al.[36] showed when radioactive morphine (^{14}C and ^{3}H-labeled) was fed *in vitro* to freshly collected samples of capsules and stem latex of *P. somniferum*, some of it was converted to radioactive N-oxide. Although metabolic activity and variation between samples of latex collected at different times were much less marked than those previously found using *in vitro* methods, the results do confirm that isolated latex is a metabolically viable tissue.

As stated earlier, Phillipson et al.[33] reported the presence of N-oxides of thebaine, codeine, and morphine in *Papaver somniferum* and *P. bracteatum*. They found low yield of the three alkaloids, N-oxides and suggested that they do not accumulate but are either transformed into other metabolites or returned to the corresponding bases. In the fresh capsule of *P. somniferum* morphine N-oxide was isolated from its more polar basic components. The major isomer of codeine N-oxide was also isolated from *P. somniferum*. These findings indicate that morphine metabolites other than normorphine are produced naturally in the plant.

The rapid turnover of morphine and related alkaloids in *P. somniferum* indicates that these alkaloids take part in metabolism and hence it is highly probable that N-oxides must also be involved in active metabolism. In *P. somniferum* latex the alkaloids are almost exclusively located in vesicles and the more polar nature of the N-oxides might ensure their exclusion or their retention in these organelles.[31,32]

In their opinion, the morphinanes represent yet another group of alkaloids which occur in the more polar N-oxide form with tertiary bases. The authors prepared N-oxides of morphine, codeine, and thebaine. Each alkaloid forms two N-oxides which have been separated and characterized by PMR,MS and reduction to the parent alkaloid. Both N-oxides of morphine and one N-oxide of codeine have been isolated from *P. somniferum* and both N-oxides of thebaine have been isolated from *P. bracteatum*[33] (Figure 9.2).

The findings of N-oxide of thebaine, a relatively minor alkaloid, codeine, and morphine in *P. bracteatum* and *P. somniferum* permits the speculation that they might be involved in active metabolism.[4]

ENZYMIC ACTIVITY

Kovacs and Benesova[14] observed the participation of aminopeptidase in aromatic amino acid formation for synthesis of alkaloid. Maximum activity of aminopeptidase occured between second and fourth day of germination.[15] Maria et al.[16] reported that during germination of poppy seed aminopeptidase activity increased up to 12 hours. Four isozymes of this enzyme were also detected.

The presence of enzymes of tricarboxylic acid cycle and glyoxylic acid cycle in the latex like aconitase and isocitrate dehydrogenase were detected by Antoun and Roberts.[39] Enzymes of glycolysis such as pyruvate kinase, lactate dehydrogenase, and those associated with peroxisomes (glyoxylate reductase, catalase) were also identified. Enzymes reported from poppy seedlings such as peroxidase, decarboxylases, etc., were also present. They also reported methyl transferase enzyme from the latex and when methionine (C^{14}-methyl) was incubated with latex, radioactive methyl group occurred in the alkaloid.[40]

The enzymes which catalyse hydroxylation, oxidative deamination, transamination, and decarboxylation of amino acids are involved in biogenesis of alkaloids in poppy plant.[41] Kovacs and Jindra[42] reported that phenolase is the enzyme which brings about hydroxylation of tyrosine. Robinson and Nagel[44] observed that extracts of *Papaver somniferum* that had peroxidase activity were ineffective in catalysing oxidation of reticuline. Two peroxidases were purified from young seedlings but only one of them was active toward indole-3-acetic acid (IAA).

FIGURE 9.2. *N*-oxides of *Papaver* alkaloids. Thebaine *N*-oxide, major isomer (1); minor isomer (2); codeine *N*-oxide, major isomer (3); minor isomer (4); morphine *N*-oxide, major isomer (5). 1 and 2 were isolated from *P. bracteatum* (thebaine-rich strain), 3, 4, and 5 were isolated from *P. somniferum* (Halle strain). Reprinted with permission from *Phytochemistry*, Vol. *15*, 1297 (Phillipson, Handa, El-Dabbas). © 1976, Courtesy of the author, Pergamon Press, and Plenum Publishing Corporation, New York.

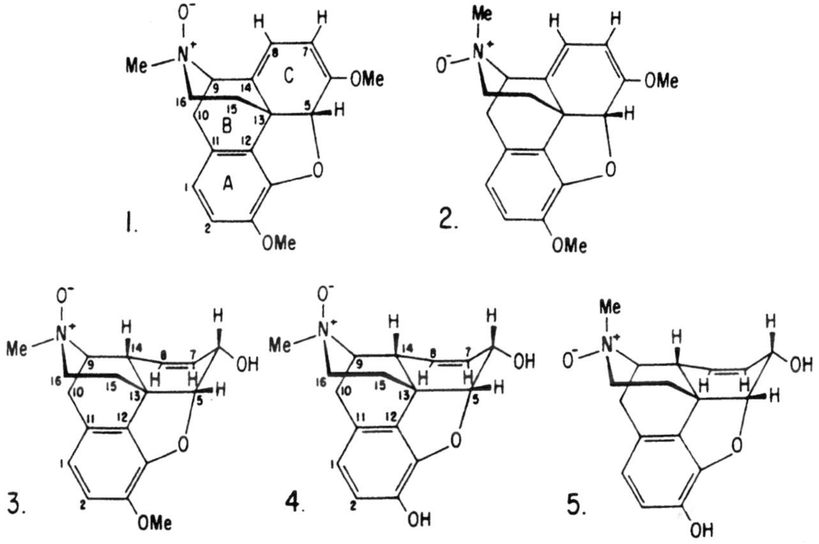

Roberts and Antoun[100] however failed to detect the necessary enzymes such as methyltransferase and L-Dopa-decarboxylase in the 1000g pallet: instead they were found in the supernatant. Even when this supernatant was further centrifuged at 4500g and 10000g the L-Dopa decarboxylase activity still remained in the supernatant.[34]

BOUND ALKALOIDS

The polysaccharide fraction of the pericarp and seed of *Papaver somniferum* were shown to contain "bound" forms of morphine which were derived from radioactive morphine fed to living plants.[45] Bound forms of codeine, thebaine, and some unidentified alkaloid-like compounds were also detected in the pericarp and bound thebaine occurred in the pericarp of *Papaver bracteatum*. The complexity and molecular weight of the bound alkaloids seemed to increase during ripening and it is suggested that these substances represent transitional forms in the metabolism and translocation of morphine from latex to seed.

METABOLITES

Morphine has been shown that it is not an "end-product" substance but at certain stages of development is rapidly metabolized, especially in the capsule. Morphine N-oxide and normorphine have both been shown to be metabolites of morphine. Miller et al.[47] suggested that the known sequence in the plant, thebaine to codeine, codeine to morphine, and morphine to normorphine which involves successive demethylation, indicates that these alkaloids act as methylating agents (Figure 9.3).

That normorphine is definitely an active metabolite of morphine in *Papaver somniferum* was shown by (a) demonstrating the presence of normorphine throughout the life cycle of the plant, (b) finding normorphine ^{14}C after feeding morphine ^{14}C via the roots, and (c) exposing opium poppies to [14]CO$_2$ under steady-state conditions which led to morphine and normorphine of the same specific gravity. Feeding normorphine ^{14}C showed that the N-demethylation step is irreversible. These results indicate that the major, if not the sole, morphine degradative pathway involves an initial demethylation to normorphine, which is subsequently degraded to non-alkaloidal

FIGURE 9.3. The conversion of morphine to normorphine. Reprinted with permission from *Phytochemistry*, 12 (Miller et al.), © 1973. Courtesy of Pergamon Press, and Plenum Publishing Corporation, New York.

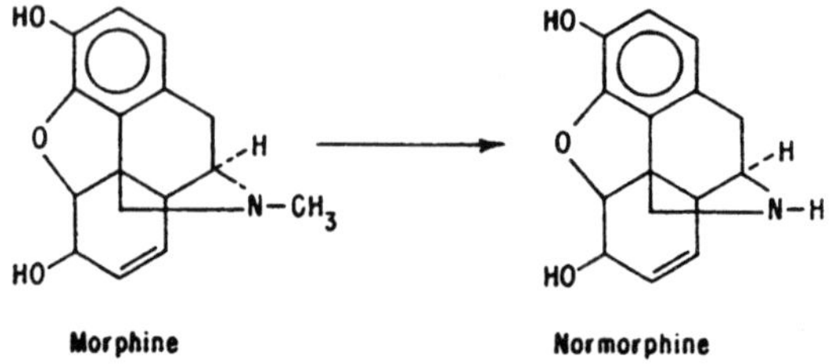

Morphine   Normorphine

metabolites. The high rates of turnover observed led to the conclusion that morphine alkaloids do play an active metabolic role, perhaps as specific methylating agents. Earlier it was suggested by Rapoport et al.[48] that both thebaine and codeine are dynamic intermediates in the plants' metabolism. Results from several $^{14}CO_2$ exposures indicated that the rate of synthesis of these morphine precursors required that they be metabolically active, since the concentration of neither increased to any extent.[47] Marked daily variations in the concentration of morphine alkaloids and morphine ^{14}C feedings suggest that morphine is broken down further to non-alkaloidal metabolites.[32,35]

BIOSYNTHESIS ACTIVITY IN PELLETS

The biosynthesis of the five major alkaloids of *Papaver somniferum* from radioactive dihydroxyphenylalanine has been studied in 1000g, 10,000g, and 100,000g pellets, and the 100,000g supernatant fractions of the capsule latex.[34] Definite evidence of biosynthesis was obtained only in the 1000g pellets. None was found in the other fractions although electron microscoy showed that organelles including vesicles were present.

A section of the 10,000g pellet shows vesicles surrounded by densely packed cytoplasm largely composed of polysomes. The vesicles were less numerous than those in the 1000g pellet and smaller, being on average about 0.5μm against about 1μm in the 1000g pellet. Most had smooth to granular walls and it has been suggested that this granular material contracts and zonates to give the familiar capped vesicles. The zonation in the cap is very distinct with an inner region different from both luman and cap. The pellet consisted mainly of large alkaloidal vesicles.

The amount of alkaloids biosynthesized, however, were very small relative to the amounts involved in the rapid changes for the developing capsule. In contrast, all fractions of the latex were able to metabolize T-morphine *in vitro* with the 100,000g supernatant showing the highest activity and the amounts involved were also consistent with the changes found in the living plant.[34]

However, ontogeny studies show that normorphine is never the most abundant alkaloid. It is known that the other major alkaloids, thebaine, codeine, and morphine, all take their turn at being the most abundant alkaloid, the sequence reflecting the biosynthetic pathways. Normorphine concentration fails to show the same behavior with age which implies that the rate of turnover is fairly constant over the life cycle of the plant. The rapid equilibration of the morphine and normorphine activity pools seen in the $^{14}CO_2$ exposure of necessity implies that active synthesis and degradation are occurring in the entire plant, possibly in stem latex. It was observed that senescent, dying plants had almost no normorphine despite the fact that morphine concentration was still quite high.[47]

Fairbairn et al.[32] showed that if radioactive tyrosine was fed to the poppy capsule, radioactive morphine was detected. Similarly, Wold et al.[43] reported formation of radioactive morphine and codeine when radioactive tyrosine was applied.

In 1959, Leete[49] obtained radioactive morphine when DL-phenylalanine-2-C^{14} or DL-Tyrosine-2-C^{14} was fed to *Papaver somniferum* plants. Systematic degradation of morphine derived from the tyrosine yielded compounds whose activities were compatible with the hypothesis that morphine is formed from two molecules of tyrosine via norlaudanosine. Rapoport et al.[48] reported the incorporation of radioactivity into each of the isolated alkaloids, morphine, codeine, and thebaine when the plants of *Papaver somniferum* were grown in the presence of $C^{14}O_2$ for different times. The incorporation of radioactivity into each of these alkaloids was determined for (1) intact compounds, (2) the various O-and N-methyl groups, and (3) ring skeleton. Differences found in the ring skeleton labeling are best accommodated by a scheme in which thebaine is the precurser of the other morphine alkaloids. Morphine appeared to be a storage product formed from codeine by demethylation. Rapoport et al. observed that the most obvious and simplest relationship which would be in accord with and explain the data obtained is that of the scheme: precurser–thebaine–codeine–morphine.

Stermitz and Rapoport[46] observed that among hydrophenanthrene alkaloids in Papaveraceae (a) thebaine is the most widely occurring and (b) thebaine is the most rapidly formed and with the highest specific activity when $^{14}CO_2$ is fed. The primary role of thebaine has been further corroborated by feeding experiments with *P. somniferum* using radioactive thebaine, codeine, and morphine in which it is found that thebaine is converted to codeine and morphine, codeine is converted to morphine, and no other conversions take place within this group. Also, in *P. orientale*, thebaine appears to be converted to oripavine. This establishes thebaine as the precurser of the other hydrophenanthrene alkaloids and O-demethylation as an important biosynthetic pathway. Comparison of specific activities of fed and recovered thebaine indicates a rapid synthesis and transformation and an important role in the plants' economy for this alkaloid. Comparative rates of $^{14}CO_2$ incorporation indicated that in *P. somniferum* thebaine (Ia) is the first hydrophenanthrene alkaloid formed and is converted by successive O-demethylations to codeine (IIa) and finally to morphine (IIb) (Figure 9.4).

This proposal was also supported by the fact that morphine, codeine, and oripavine (Ib) are found in a number of species of Papaveraceae but only together with thebaine, while the converse is not true. These relationships are shown in Table 9.3.

[a] Asahima, T. Kawatani, M. Duo, and S. Fujita, *Bull. Narcotics, U.N. Dept. Social Affairs, 7*, No. 2, 20 (1957). [b] R. A. Konovalova, C. Yunusov, and A. P. Orekhov, Ber., 68B, 2158 (1935). [c] R. A. Konovalova and U. V. Kiselev, *Zhur. Obshchei Chim.*, 18, 855 (1948). [d] F. Santavy, M. Maturova, A. Nemeckova, H. B. Schroter, H. Potesilova, and Vl. Preininger, *Planta Med.*, 8, 167 (1960).

Waller and Nowacki[4] observed that some biosynthetic pathways are restricted to smaller units, i.e., to sections in a genus. Such is the

FIGURE 9.4. Conversion of thebaine to morphine. Reprinted with permission from *J. Amer. Chem. Soc.*, *83* (Stermitz & Rapoport). © 1961. Amer. Chem. Soc.

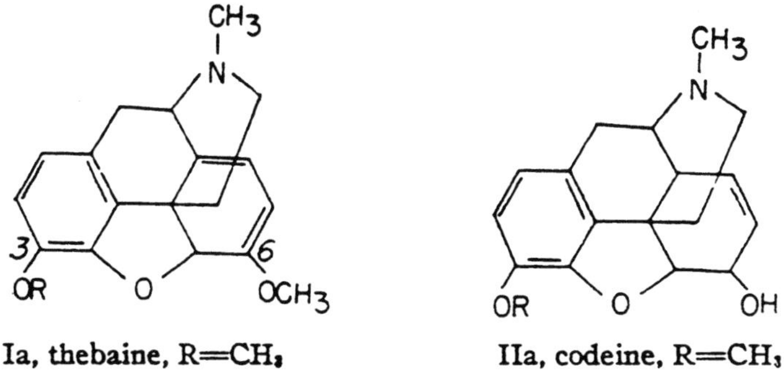

Ia, thebaine, R=CH₃
Ib, oripavine, R=H

IIa, codeine, R=CH₃
IIb, morphine, R=H

TABLE 9.3. Opium Alkaloids in Papaveraceae

Species	Morphine	Codeine	Oripavine	Thebaine
P. somniferum L.	+	+	−	+
P. setigerum D.C.[a]	+	+	−	+
P. orientale L.[b]	−	−	+	+
P. bracteatum Lindle.[c]	−	−	+	+
P. strigosum Schur.[d]	−	−	−	+
P. intermedium O. Ktze.[d]	−	−	−	+

Reprinted with permission from *J. Amer. Chem. Soc., 83,* 4045 (Stermitz and Rapoport) © 1961, Amer. Chem. Soc.

case for the morphine type of alkaloids which are only found in the section Mecones in the two species *Papaver somniferum* and *P. setigerum.* The closely related *P. orientale* and *P. bracteatum* belong to section Macrantha and they accumulate either thebaine, a substrate for morphine, or isothebaine, which can be demethylated to give oripavine, an isomer of codeine. Both species of Macrantha do hybridize with *P. somniferum* giving an infertile F_1 generation.

The section Mecones in *Papaver* has only five surviving species and the section Macrantha only four. These sections are closely related, yet they are difficult to cross and the cross usually produces only sterile hybrids. The differences in alkaloids is small but remarkable. The Mecones accumulate codeine and morphine while Macrantha transformation stops with the formation of thebaine or its homologue isothebaine. The Mecones differ from Macrantha by an additional character: the main alkaloids are all derivatives of (+)–reticuline and at least two pathways of alkaloid conversion from norlaudanosine in *P. orientale* operate. One proceeds through reticuline, the other through orientaline.

It has been observed that Papaver species are capable of producing thebaine, some in very high yield but only two, *P. somniferum* and *P. setigerm,* appear to be able to generate morphine. Most other thebaine-producing plants lack the enzyme required for demethylation of the enol ether. Nonetheless, several *Papaver* species which contain thebaine are capable of 3-O-demethylation of the phenol ether to oripavine, which appears to represent the last step in the biosynthesis of morphinan alkaloids in these plants.[61]

Martin et al.[50] reported the natural occurrence of diphenolic benzyltetrahydroisoquinoline alkaloid, reticuline, in fresh-budding plants and seedlings of *Papaver somniferum*. Exposure of such plants to $^{14}CO_2$ for one to three hours was followed by determination of radioactivity incorporated: (a) into reticuline and thebaine and (b) into the N– and O– methyl groups for both alkaloids. These results confirm those from feeding experiments and establish beyond question the role of reticuline as the true biosynthetic benzyltetrahydroisoquinoline precursor of the hydrophenanthrene alkaloids. The high rate of incorporation of radioactivity into the total alkaloids as well as into thebaine and reticuline both in seedlings and in mature plants should place beyond doubt the intimate involvement of these alkaloids in the economy of plant. According to the authors, these findings constitute positive evidence for the operation of the same biosynthetic relationships (i.e., $CO_2 \rightarrow$ reticuline $\rightarrow$ thebaine $\rightarrow$ codeine $\rightarrow$ morphine) in seedlings as in the mature plant. This confirms the earlier works[46,48] wherein rapid *de nova* synthesis of thebaine from $^{14}CO_2$ and its primacy in the hydrophenanthrene alkaloid series is established.

Horn et al.[51] conveyed that biosynthesis of Morphinan alkaloids proceeds by conversion of the enol ether of thebaine to the keto group of neopinone and thence to codeinone. To determine the mechanism of this transformation [G-^{14}C,6-^{18}O], thebaine was fed to *Papaver somniferum* and the codeine and morphine were isolated. Comparison of the $^{18}O/^{14}C$ ratios in the codeine and morphine isolated with that of the thebaine fed showed that about 34 percent of the ^{18}O had been retained. Parallel feedings with [G-^{14}C,6-^{18}O] codeinone demonstrated that the loss was due to nonenzymic exchange. Thus, the mechanism of enol ether cleavage in thebaine is established as cleavage of the 6-O-methyl group with retention of the 6-oxygen in the codeinone.

The biosynthetic pathway for the opium alkaloids in *Papaver somniferum* has been shown to proceed by the conversion of thebaine (1) to codeinone (3) probably via neopione (2) and migration of the double bond into conjugation. Codeinone (3) is then reduced to codeine (4) which subsequently is 3-0-demethylated to morphine (5) Morphine in turn is N-demethylated to normorphine (6) and the metabolic fate of the latter has yet to be determined. Thus, in a formal sense O-demethylations occur at 2 steps in the pathway: (a) the conversion of the thebain (1) to codeinone (3) and (b) the conversion of codeine to morphine (Figure 9.5).

The authors hypothesize that these two reactions may be catalyzed by different types of enzymes. Since reaction (a) involves an enol ether and reaction (b) an aromatic ether, different mechanisms may be involved in their cleavage, as is the case *in vitro*. Also, morphine and codeine are peculiar to *P. somniferum* while thebaine occurs in most other species as well particularly in *P. orientale* and *P. bracteatum*.[46] Oripavine (7) also occurs in these two species. These data strongly support the hypothesis that all the three species contain an O-demethylase which is capable of cleaving the aromatic ether linkage, while *P. somniferum* is unique in that it contains an enzyme system which attacks the enolic ether as well.

Hodges and Rapoport[54] observed that although the identification of the intermediates in the pathway of morphine biosynthesis in *P. somniferum* seems nearly complete, little progress has been made in the study of enzymic systems which catalyze the individual steps in the biosynthesis of the alkaloids. Even detection of enzymes responsible has been difficult outside the living plant.

To provide a basis for isolating the enzymes responsible for alkaloid biosynthesis the authors have utilized cell-free extracts from entire opium plants to demonstrate the specific *in vitro* conversion of codeinone (1) to codeine (2). They developed a method of quantitating this conversion at the n mol level to allow the possibility of preparing active enzyme extracts.

They prepared cell-free extracts from *P. somniferum* which catalyze the reduction of codeinone [16-^{3}H] to codeine. The methodology for examing this conversion has pointed to conditions for exploring the preparation of suitable enzyme extracts. *P. bracteatum* also yielded a cell-free system which reduced codeinone to codeine, both of which are foreign to this species (Figure 9.6).

Antoun and Roberts[52] reported that the formation of the hydrophenanthrene, phthalideisoquinoline, and benzylisoquinoline groups

FIGURE 9.5. Biosynthesis of morphinan alkaloids. 1. thebaine; 2. neopione; 3. codeinone; 4. codeine; 5. morphine. Reprinted with permission from *J. Am. Chem. Soc.*, 100, 1195 (Horn, Paul, and Rapoport). © 1978. Amer. Chem. Soc.

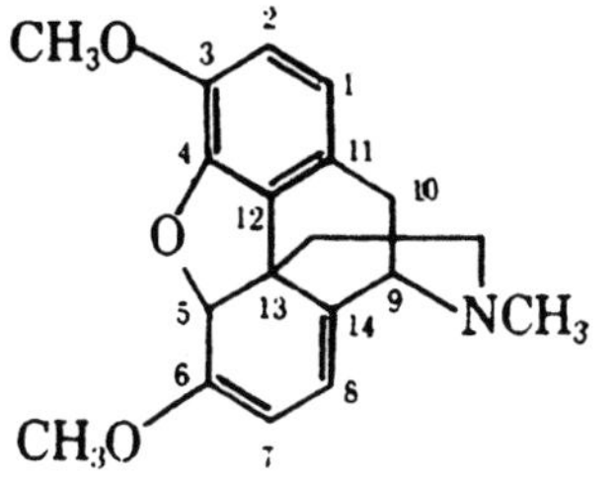

1

→

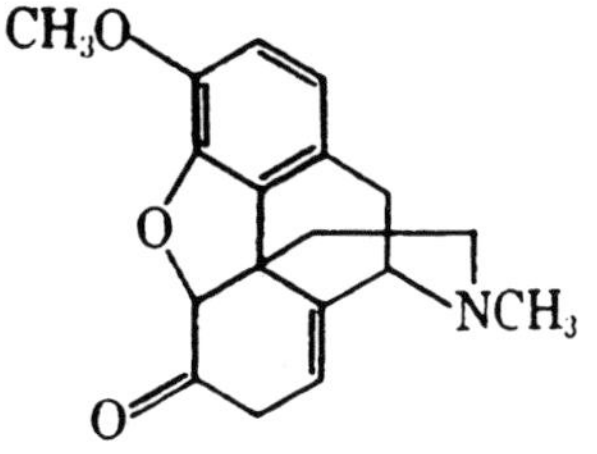

2

→

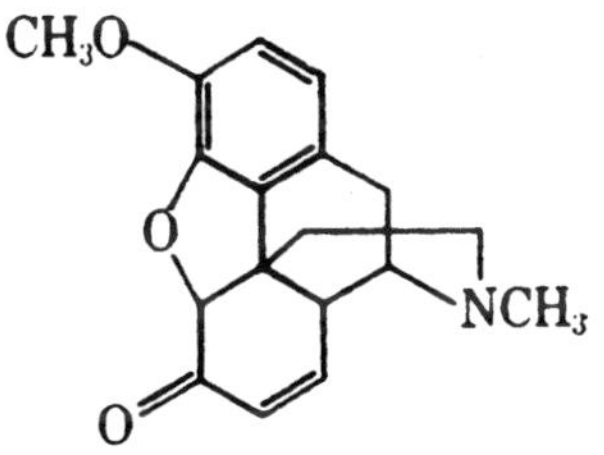

3

→

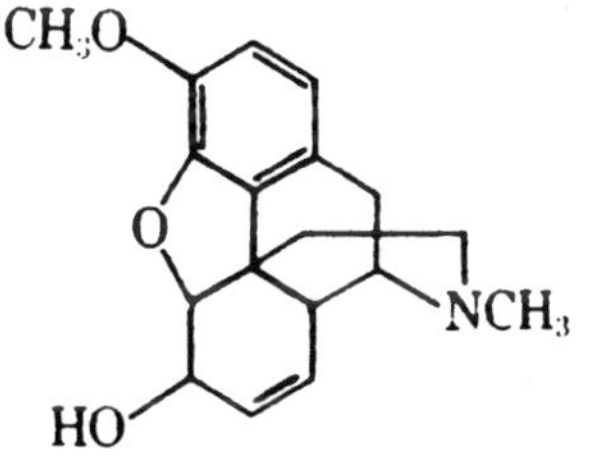

4

→

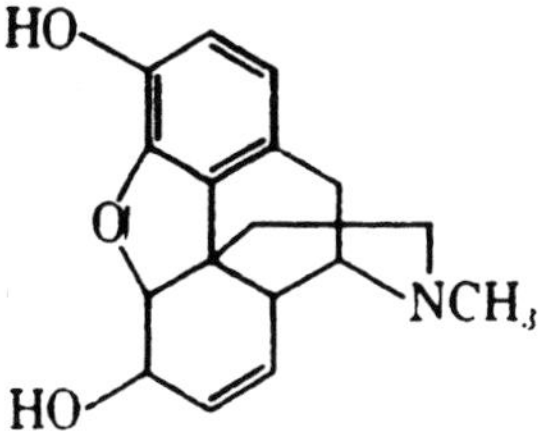

5

→

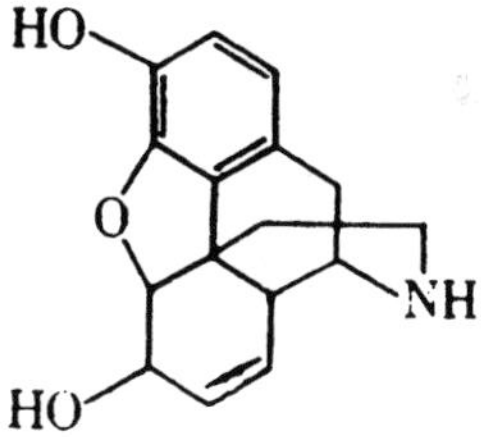

6

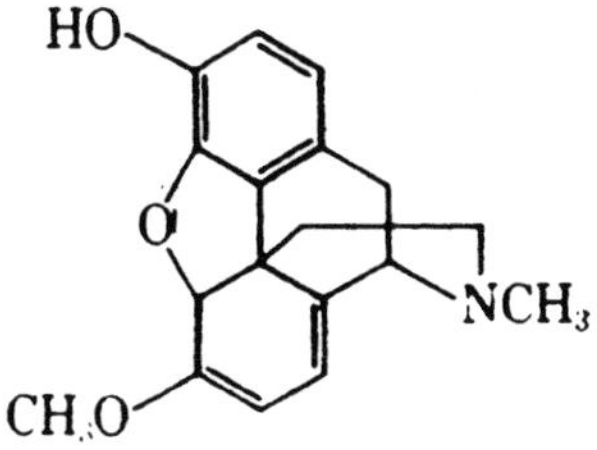

7

FIGURE 9.6. Conversion of Codeinone to Codeine. 1. codeinone; 2. codeine. Reprinted with permission from *Phytochemistry* 19, 1681, 1980 (Hodges and Rapoport). © 1980. Pergamon Press.

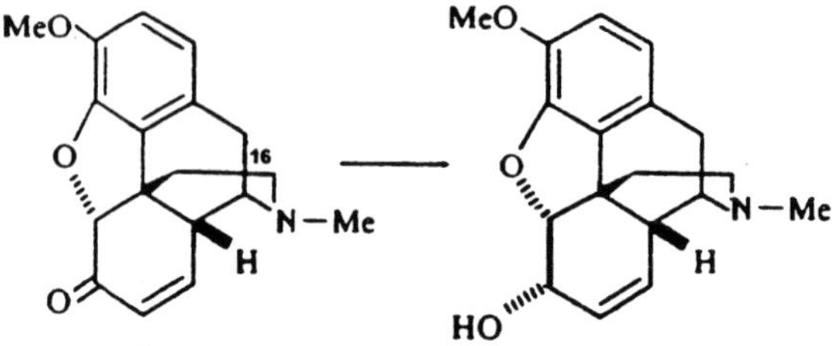

of alkaloids involves a number of methylation and demethylation steps. Spencer[53] indicated that methionine was the most efficient donor for both O- and N-methyle groups of the alkaloids. The preliminary study indicated that poppy latex was capable not only of carrying out the necessary methylation reactions of the alkaloids but also capable of demethylating thebaine to codeine and possibly morphine. The presence of methylating enzymes in the whole latex, as well as latex fractionated to 1000g × 30 min. and supernatant fractions was detected. Radioactive reticuline, codeine, thebaine, and papaverine were found.

Narcotine was not present, indicating either that the enzymes involved in its formation were not active at the time of collection of latex or that its biosynthesis occurs outside the laticiferous vessels.[52]

Williams and Ellis,[57] while studying the accumulation of alkaloids in tissues of *P. somniferum* cv. Marianne from germination to post-petal-drop stage, observed the highest accumulation in day 30 root tissue where morphine and narcotoline reached the concentrations of 313 µg/g and 490 µg/g fresh weight, respectively. In their opinion juvenile root tissue may be a useful tissue source for studies of the biosynthesis of these alkaloids. Morphinan and phthalideineisoquinoline alkaloids were found to accumulate in a time- and tissue-specific manner.

BIOTRANSFORMATION OF ALKALOIDS

Furuya et al.[55] reported that callus tissue of *Papaver somniferum* contains benzophenanthridine, protophine, and aporphine-type alkaloids, although the opium poppy contains the main components of morphinan-type alkaloids. This fact suggests that the callus tissue cannot biosynthesize the alkaloids derived from (–)–(R)-reticuline, an important intermediate of opium alkaloids such as morphine but can biosynthesize only those from (+)–(S)–reticuline. The authors converted (RS)-reticuline stereospecifically to (–)–(S)-scoulerine and (–)–(S)-cheilanthifoline by cell cultures of *P. somniferum* and (–)–(R)–reticuline was recovered as an optical pure compound by racemic resolution. (–)–Codeinone was converted in high yield to (–)–codeine in both cell culture and enzyme preparation, but the other morphinans, thebaine, codeine, and morphine, were not metabolized.

Tam et al.[56] reported that thebaine is biotransformed to neopine by cell suspension cultures of *Papaver somniferum* cv. Marianne grown in 0-B5 medium. While studying the effect of precursors on the cell suspension cultures of *P. somniferum* cv. Marianne only thebaine and codeinone gave rise to neopine and codeine, respectively. The fact that codeine was not metabolized to morphine whereas codeinone was transformed to codeine indicates that this cell line lacks the ability to demethylate codeine to morphine. Conversion of thebaine to neopine instead of neopinone indicates that transformation in plant cell cultures does not necessarily follow the biosynthesis pathway for conversion of thebaine to neopinone in opium poppy as established by Parker et al.[111]

Robinson and Nagel[44] observed that *peroxidase* activity present in the extracts of *P. somniferum* was ineffective in catalysing oxidation of reticuline. Out of two peroxidases purified from young seedlings, only one was active toward indole-3-acetic acid (IAA). Both young seedlings and latex of the mature poppy plants, in their opinion, synthesize morphinan compounds from the precurser (–) reticuline [1] and the first identified product is (–) salutaridine [2]. The authors observed that peroxidase was present in seedlings at a time when morphinan alkaloids are being synthesized but characterization of them did not indicate any role in alkaloid formation (Figure 9.7).

In 1971 Roberts[58] had reported that enzymes of either peroxidase or laccase type appear to be likely candidates for catalysing such a reaction, but no peroxidase activity has been found in latex of

FIGURE 9.7.1. (–) Reticuline; 2. (–) Salutaridine. Reprinted with permission from *Phytochemistry* 21(3) 535, 1982 (Robinson and Nagel). © 1982 Pergamon Press.

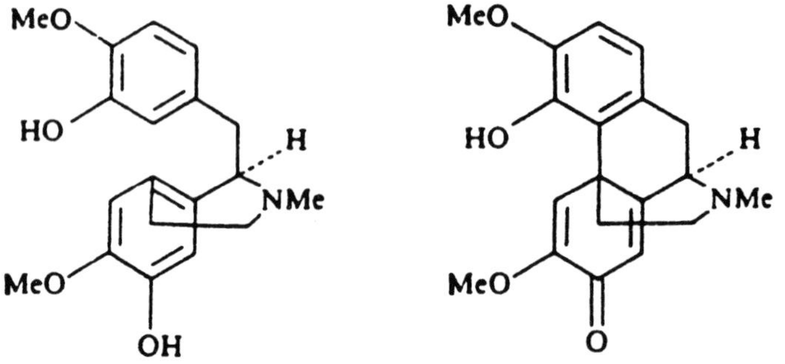

P. somniferum and phenol oxidizing preparations from this source are inactive toward (–) reticuline.

Kamo and Mahlberg[78] observed that latex and cell-free extracts of various organs and stages of plant and capsule development in *P. somniferum*, synthesized dopamine, an alkaloid precursor from ^{14}C-dopa. The 1000g x 30 min. supernatant from latex of the pedicel-capsule junction converted more dopa than latex supernatant from the upper capsule or lower pedicel regions, although there were more proteins in the latex from the capsule. Percent conversion of pedicel-capsule latex into dopamine was maximum in unopened flower buds and decreased within 14 days after flowering. Dopamine biosynthesis in latex and cell-free extracts also varied with the stage of organ development. Extracts from capsule tissue converted more labeled dopa into dopamine than did extracts from pedicel, leaves from vegetative plants at the rosette stage, leaves from flowering plants, or pedicels connected to capsules.

Hsu et al.[59] studied biotransformation of codeine to morphine in isolated capsules of *Papaver somniferum*. They found that cofactors such as nicotinamide adenine dinucleotide, adenosine 5'-triphosphate, S-acetyl coenzyme A, and pyridoxal phosphate were not required in the conversion of codeine to morphine. Reducing agents such as dithiothreitol, putathione, and β-mercaptoethanol strongly promoted codeine and morphine degradation, while morphine formation remained at a constant level. Hydrogen peroxide (concentration > 0.25 mM) caused the conversion of codeine and morphine to N-oxides by non-enzymatic oxidation. In their opinion, isolated capsules system provides an ideal method for studying the biotransformation of morphine alkaloids.

In another study Hsu and Pack[112] investigated the ability of codeine to convert to morphine in cultured cells of *P. somniferum* by radioisotope technique. Results suggest that enzymes involved in the oxidative degradation of codeine dominate the biotransformation of codeine to morphine in the cultured cells. Extracts from the cultured cells contained thebaine as the major morphinan alkaloid and small amounts of codeine, while morphine was not detected. Addition of ^{14}C-codeine to the cultured cell demonstrated that ^{14}C-codeine was predominantly converted to N-oxide products along with the formation of minor amounts of morphine indicating that induced poppy cell cultures could synthesize morphine with low efficiency.[112]

As stated earlier, latex or whole plants of *Papaver somniferum* have the ability to convert morphine and codeine into their N-oxides.[33,36] Demethylation of morphine in *P. somniferum* has been established as an active metabolic process.[47] Normorphine was found in raw opium.[33] The morphine degradative pathway involves an initial demethylation to normorphine which is subsequently degraded to non-alkaloidal metabolites. It is suggested that morphine alkaloids may play an active metabolic role, perhaps as specific methylating agents.

Brochmann-Hanssen[61] reviewed the research work done on biosynthesis of morphinan alkaloids and aptly remarked that its guiding principle was the relationship of morphine to 1-benzylisoquinolines, as first pointed out by Gulland and Robinson,[62] who put forth the correct structure of morphine based on this biogenetic approach. They proposed that morphine and related alkaloids are generated in the plant by oxidative coupling of a suitable benzyltetrahydroisoquinoline precursor such as nor-laudansoline. Earlier Winterstein and Trier[63] had suggested that the benzylisoquinoline system in nature is built up from two units derived from 3,4-dihydroxyphenylalanine (dopa). Since dopa comes from tyrosine, the early inves-

tigations consisted of feeding experiments with radioactively labeled tyrosine, dopamine, and norlaudanosoline and isolation of
radioactive morphine, codeine, and thebaine. When specifically labelled precursors were used, the label appeared in the morphine
alkaloids in the positions predicted according to hypothesis. It follows therefore that morphine is biosynthesized from two phenethyl
residues, both being derived from tyrosine and that norlaudanosoline is a likely intermediate.

Based on the work of Pummerer et al.[64] on oxidation of phenols
to produce relatively stable radicals, Barton and Cohen[65] expanded
this principle to encompass the biogenesis of alkaloidal ring systems by intramolecular coupling of phenol radicals.

ROLE OF RETICULINE

As observed by Brochmann-Hanssen, tracer experiments had
clearly demonstrated that thebaine is the first morphine-type alkaloid produced in the opium poppy and that it is converted to codeine
(4) which, in turn, is demethylated to morphine (1).[66,48,46] It was,
therefore, reasonable to expect that a precursor closer to thebaine
than norlaudanosoline should have an N-methyl group and two
O-methyl groups in the molecule. A new benzylisoquinoline alkaloid isolated from *Annona reticulata* and named reticuline (5) filled
this requirement.[67] More important, reticuline had the phenolic
characteristics and the substitution pattern suitable for intramolecular ortho-para coupling according to Barton and Cohen.[65,66] It is
interesting to point out that what may be a principal alkaloid in one
plant (reticuline in *Annona*) is, in another, a transient metabolite,
which is essential to some metabolic pathway but which does not
accumulate. Feeding experiments with labeled reticuline gave better incorporation into morphine, codeine, and thebaine than had
been achieved with either tyrosine or norlaudanosoline.[68,69] Again,
there had been no scrambling of the radioactive label. N-Norreticu-
line was also well incorporated but somewhat less efficiently than
reticuline indicating that O-methylation precedes N-methylation.
When all three methyl groups of reticuline (5) were labeled with
^{14}C in addition to a reference label in the carbon skeleton (C-3), all
labels appeared in thebaine with about the same relative activities as

in the precursor. Therefore, the biotransformations from reticuline to thebaine did not produce any change in the methylation pattern (Figure 9.8).

The idea of oxidative coupling of reticuline or a reticuline-like substance to a dienone proved to be correct. By an interesting coincidence, this morphinandienone (7) had been isolated from *Croton salutaris* and named salutaridine by Barnes as cited by Barton et al.[69] Labeled salutaridine showed good incorporation into morphine, codeine, and thebaine, and its natural occurrence in the opium poppy was demonstrated by an isotope dilution experiment.[69] Recently the oxidative coupling of reticuline to salutaridine was carried out in cell-free systems prepared from the opium poppy.[70] Ginsberg proposed that salutaridine was reduced to a dienol prior to forming the 4,5-epoxybridge.[71,72] This reduction, when carried out *in vitro* with sodium borohydride, gave a mixture of two epimers named salutaridinol I and salutaridinol II. Although both epimers could easily be converted to thebaine *in vitro*, feeding experiments demonstrated that only salutaridinol I had the correct stereochemistry at C-7 for *in vivo* conversion to thebaine by allylic elimination.[69,73]

FIGURE 9.8. Morphinan alkaloids.1. morphine; 2. norlaudanosoline; 3. thebaine; 4. codeine; 5. reticuline; 6,7. morphinandienone; 8. salutaridinol. By courtesy of the author, E. Brochmann-Hanssen in *The Chemistry and Biology of Isoquinoline Alkaloids*, Phillipson et al. (eds.), 1985. Heidelberg: Springer-Verlag.

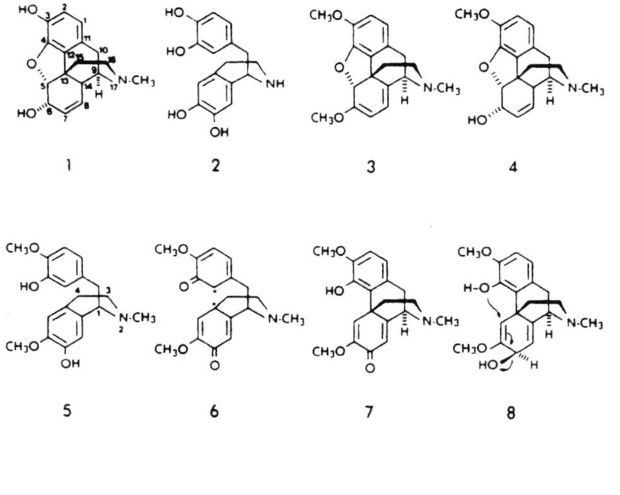

Until this time, all tracer experiments on *P. somniferum* had been carried out with racemic mixtures. When it was realized that natural S(+)-reticuline (9) isolated from *Annona reticulata* had the wrong stereochemistry relative to the morphine alkaloids, Battersby et al.[74] decided to work with the optical isomers. When (+) and (–)–reticuline, carrying a ^{3}H label at the asymmetric center (C-1) and ^{14}C labels at three other positions in the molecule, were fed to separate batches of opium poppies, both enantiomers were incorporated into the morphine alkaloids with about equal efficiency but with loss of [3]H. The ^{3}H loss was almost 100 percent when (+)–reticuline was the precursor but only 60-80 percent from (–)–reticuline feedings. These surprising results were accommodated by proposing a reversible oxidation-reduction system via 1,2-dehydroreticulinium ion (10) accounting for the loss of ^{3}H and permitting inversion of configuration at C-1 (Figure 9.9).

Indeed, the feeding of labeled 1,2-dehydroreticulinium chloride gave the best incorporation into morphinan alkaloids achieved up to that time. Later the natural occurrence of this substance was demonstrated in the opium poppy.[76] Prior to the feeding experiments with optically active reticuline, this alkaloid had been isolated from opium and found to be present in both enantiomeric forms with an excess of the S(+)–isomer over the R(–)–isomer by Brochmann-Hanssen et al.[76,77] This was confirmed by isotope dilution studies on the fresh plant. In the seedlings, where the biosynthesis of morphine is not active, both isomers were found to be present in about equal amounts. Since then, reticuline has been isolated from many plants belonging to several plant families and has been found to play an important role in the biosynthesis of a large number of alkaloids.

BIOCONVERSION OF THEBAINE TO MORPHINE

The transformation of thebaine (3) had been considered on chemical grounds to involve hydrolysis of the enol ether to give neopinone (11) which, after isomerization to codeinone (12), could be reduced to codeine[65] (4) (Figure 9.10). This hypothesis was tested independently by two research groups and found to be essentially correct.[79,80,81] Codeinone and neopinone were excellent precursors

FIGURE 9.9. Interconversion of S and R reticuline. 9. (+) reticuline, 10. 1,2-dehydroreticulinium ion. Courtesy of the author, E. Brochmann-Hanssen, 1985.

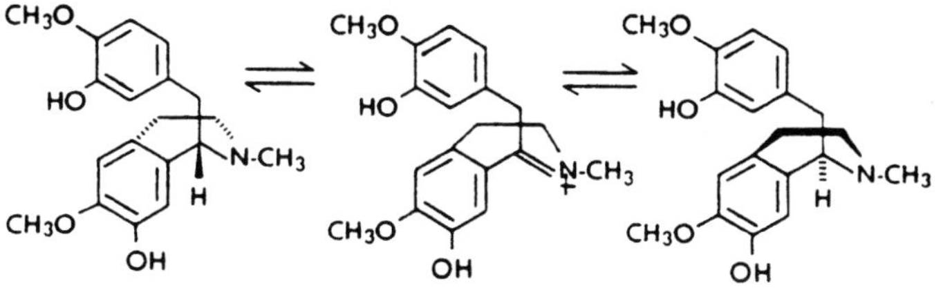

9   10

FIGURE 9.10. Biotransformation of thebaine to morphine via codeinone and codeine. 3. thebaine; 11. neopinone; 12. codeinone; 4. codeine; 1. morphine; 13. neopine. Courtesy of the author, E. Brochmann-Hanssen, 1985.

3   11   12

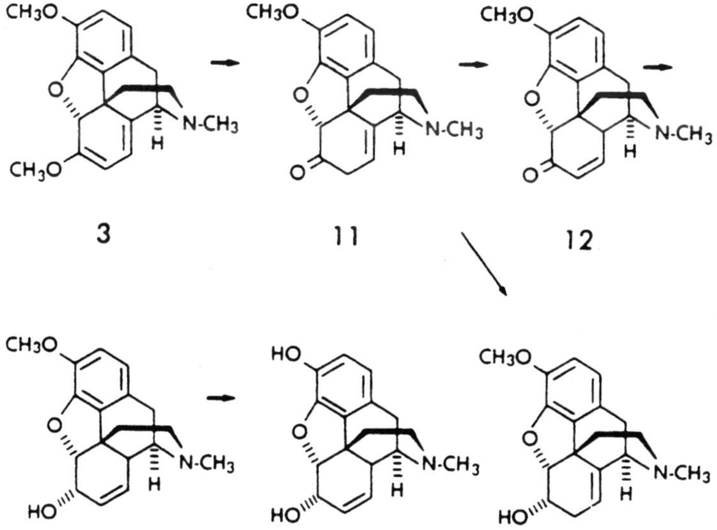

4   1   13

of codeine and morphine and could be detected in the plants by carrier dilution after exposure to $^{14}CO_2$. The intermediary neopinone also explained the presence of neopine (13) which had been isolated from opium.[82]

Many *Papaver* species are capable of producing thebaine, some in very high yield, but only two, *P. somniferum* and *P. setigerum*, appear to be able to generate morphine. Most other thebaine-pro-ducing plants lack the enzyme required for demethylation of the enol ether. Nonetheless, several *Papaver* species which contain thebaine are capable of 3-O-demethylation of the phenol ether to oripavine (14), which appears to represent the last step in the bio-synthesis of morphinan alkaloids in these plants. In the opium poppy 6-O-demethylation of the enol ether of thebaine and 3-O-de-methylation of the phenol ether of codeine proceed by the same oxidative mechanism[51] but appear to require different enzymes although these enzymes show relatively little substrate specificity. Major structural modifications of codeine and thebaine do not affect the efficiency of the dealkylations to any great extent.[83,84,85] This lack of specificity led to speculations as to the presence of oripavine in *P. somniferum*. Nielson et al.[86] isolated oripavine from the dried capsules of opium poppies cultivated in Tasmania and was later shown to be present in other chemical strains as well (Figure 9.11).

This raised a question as to the metabolism of oripavine in the plant and its possible role in the biosynthesis of morphine. When [2-³H]oripavine was fed to two different strains of *P. somniferum*, very high incorporation was achieved into morphine (1) and also into morphinone (15) which was added as cold carrier during the extraction.[87] Radioactive oripavine was isolated from both chemi-cal strains in larger amounts but of much lower specific activity than the precursor which had been fed, indicating dilution with a natural pool of oripavine. Thebaine and codeine were not radioac-tive. Therefore, demethylation of the phenolic ether of thebaine was not reversible. These results established the sequence: thebaine (3) → oripavine (14) → morphinone (15) → morphine (1) as a second

FIGURE 9.11. Biotransformation of thebaine to morphine via oripavine and mor-phinone. 3. thebaine; 14. oripavine; 15. morphinone; 1. morphine. Courtesy of the author, E. Brochmann-Hanssen, 1985.

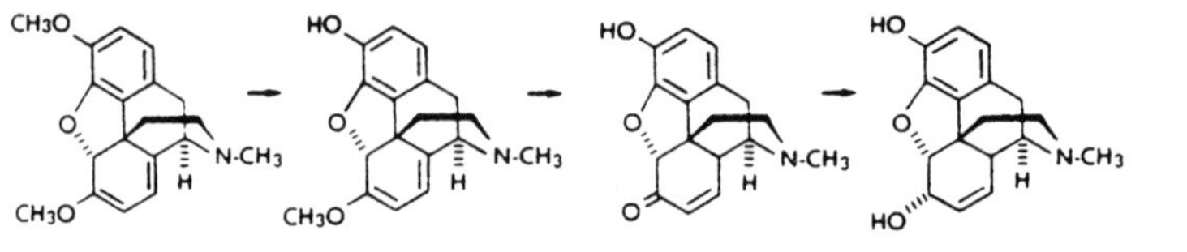

3   14   15   1

pathway for the conversion of thebaine to morphine (Figure 9.11). The close agreements and the high incorporation achieved with two chemical strains of *P. somniferum* as well as the presence of codeine and oripavine in both suggested that both pathways were operating simultaneously. The predominance of one pathway over the other may depend on the relative activities of 3-O-methyl oxidase and 6-O-methyl oxidase acting on thebaine.

The biosynthesis of thebaine in *P. bracteatum* follows the same pathway as in the opium poppy.[88,89] In contrast to earlier studies, recent isolation work has revealed that *P. bracteatum* contains small amounts of codeine, neopine, and oripavine as well as 14-β-hydroxycodeinone (16) and 14-β-hydroxycodeine (17) and both isomers of thebaine N-oxide.[90,33,91,92] It appears that the O-methyl oxidases in *P. bracteatum* are extremely sluggish in comparison with the corresponding enzymes in *P. somniferum*. This leads to accumulation of thebaine which is slowly and inefficiently converted to oripavine and to neopine and codeine. The unusual 14-β-hydroxyderivatives (16,17) may be derived from thebaine via an oxygenated intermediate as illustrated in Figure 9.12.

METABOLISM AND IN VIVO DEGRADATION OF MORPHINANS

The natural pools of alkaloids change and fluctuate not only as a result of the age and development of the plant but also on a daily and even hourly basis.[93,30,94,95] Miller et al.[47] found that morphine was N-demethylated to normorphine in the opium poppy. This demethylation, like all other biosynthetic reactions preceding it, was

FIGURE 9.12. Proposed biosynthesis of 14-β-hydroxycodeinone and 14-β-hydroxycodeine in *P. bracteatum*. 5. Reticuline; 16-14B. hydroxycodemone; 17-14B. hydroxycodeine. (Courtesy of the author, E. Brochmann-Hanssen, 1985.)

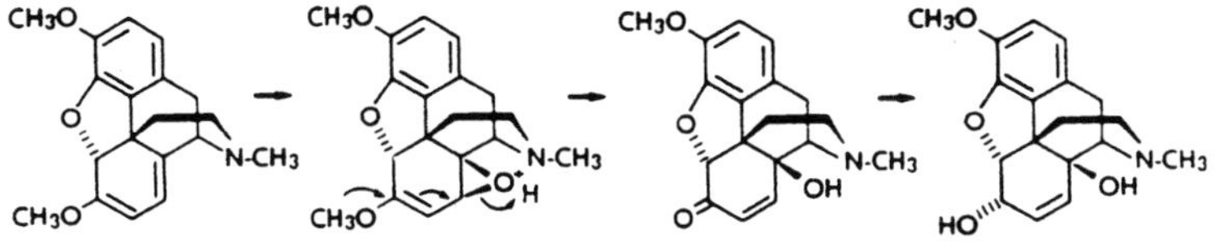

5   16   17

irreversible and its rate was fairly constant over the life of the plant. They suggested that the alkaloids may serve a biochemical function as specific methylating agents. The N-oxides of morphine, codeine, and thebaine have been isolated, but their function in the plant is not yet understood.[33] They may represent just another degradative mechanism or they may, because of their polar nature, play a role in the translocation of the alkaloids or be implicated in the general oxidation-reduction processes of the plant.

Although morphine becomes the major alkaloid at the height of the plant's development, it is clearly not an end-product in this secondary nitrogen metabolism. Fairbairn and his coworkers demonstrated a fairly rapid metabolism of morphine to products which increased in complexity and molecular weight during the ripening of the plant. They suggested that these products may represent transitional forms in the metabolism and translocation of the alkaloids from the latex to the ovules and the seeds.[35,25,28,45] Morphine bound to water-soluble high molecular weight carbohydrates were found in the capsules of *P. somniferum*.[96] Whether these various products derived from morphine and other morphinan alkaloids play active roles with specific functions or whether they represent metabolic detoxification products which are channeled back into the plant's economy is still open to question.

STEREOSELECTIVITY AND SUBSTRATE SPECIFICITY

The vast majority of biosynthetic reactions are enzyme-catalyzed and, therefore, show considerable stereoselectivity and substrate specificity. In general, stereoselectivity is more important than substrate specificity, and many unnatural morphinan derivatives have been produced biosynthetically.[83,84,85,97,98] The oxidative coupling of reticuline to salutaridine as well as the succeeding biotransformations are not very substrate specific. Several reticuline analogues have been converted to the corresponding morphine and codeine analogues. On the other hand, a reaction which exhibits exceptional substrate specificity is the interconversion of the S and R isomers of reticuline, a reaction which is crucial for the production of morphine alkaloids in the opium poppy. No other natural or synthetic benzyltetrahydroisoquinoline has been found to undergo this reaction.[99,98]

It is noteworthy, observed Brochmann-Hanssen, that the early steps in the biosynthesis leading to reticuline involve a series of methylations which are stereoselective for the L(S)-configuration. The later steps from thebaine to normorphine consist of stepwise demethylation of compounds having the R-configuration at the equivalent asymmetric center.

As already reported, labeled reticuline and norreticuline both proved to be very efficient precursors of morphine in the poppy, surpassing norlaudanosoline in this respect. When tetrahydropapaverine (all hydroxyl groups methylated) was fed by injection to the opium poppy, negligible incorporation into the alkaloids was obtained. The stages in the conversion of reticuline to thebaine and of thebaine to codeine were demonstrated by the feeding of appropriate labeled intermediates, alkaloids which have since been isolated as minor components of the opium alkaloid mixture. The principal features of the biosynthesis of thebaine, codeine, and morphine as now envisaged are shown in Figure 9.13.

Schutte and Liebisch[115] quoted the example of straightforward phenol coupling of a benzylisoquinoline in the biosynthesis of thebaine, codeine, and morphine. As can be seen (Figure 9.14), reticuline present in *P. somniferum* is the substrate for the coupling step which takes place *orth-para* to give the dienone salutaridine in which only one of the two rings can rearomatize by enolization. Because reticuline is a precursor, its progenitors tyrosine, dopamine, norlandanosoline-1-carboxylic acid, and norlandanosoline are specifically incorporated into the opium alkaloids. Reticuline is produced from tyrosine, has the (S)-configuration and is dextrorotatory. It undergoes racemization in the plant, a reaction essential for the formation of the morphine alkaloids derived from R-reticuline. This racemization proceeds by a reversible oxidation reduction mechanism via the 1,2-dehydroreticulinium ion and is enzymatic and substrate specific. The presence of salutaridine in *P. somniferum* has been shown by trapping experiments. In the further biosynthetic pathway to morphinan alkaloids, salutaridine is reduced to the dienol salutaridinol-1, which undergoes a ring closure, yielding thebaine. Salutaridine and salutaridinol were effeciently incorporated into thebaine, codeine, and morphine.

In their opinion, the final steps of the biosynthesis of morphine

FIGURE 9.13. Biogenesis of thebaine, codeine, and morphine. Courtesy of the author, W. C. Evans, (*Pharmacognosy,* 12th ed., 1983.)

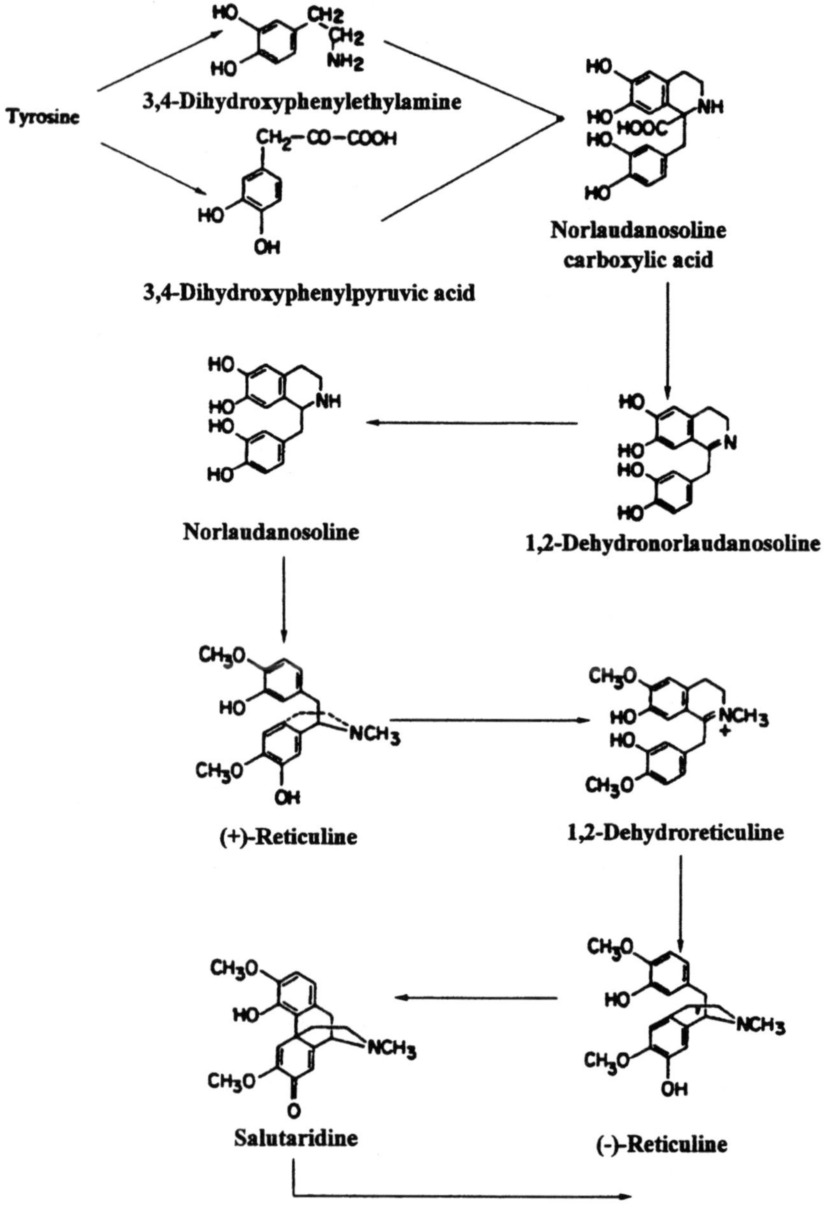

FIGURE 9.13 (continued)

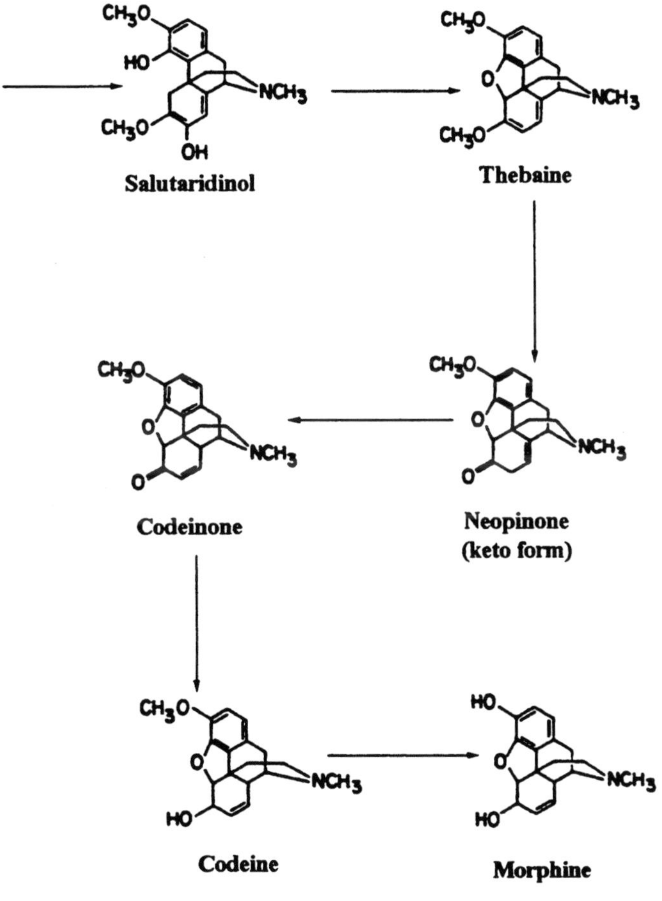

Salutaridinol

Thebaine

Codeinone

Neopinone (keto form)

Codeine

Morphine

by *P. somniferum* have been shown to involve the conversion of thebaine into codeine, followed by 3-O-demethylation of codeine to morphine. This sequence was determined using $^{14}CO_2$ exposure and precurser feeding which also demonstrated that both of these steps are irreversible. The first step proceeds by the cleavage of the

FIGURE 9.14. Biosynthesis and metabolism of morphinan alkaloids. In *Biochemistry of Alkaloids*, Mothes, Schute, and Luckner (eds.), 1985. Courtesy of Huthig Verlagsgemeinschaft, Decker & Muller GMGH, Heidelberg.

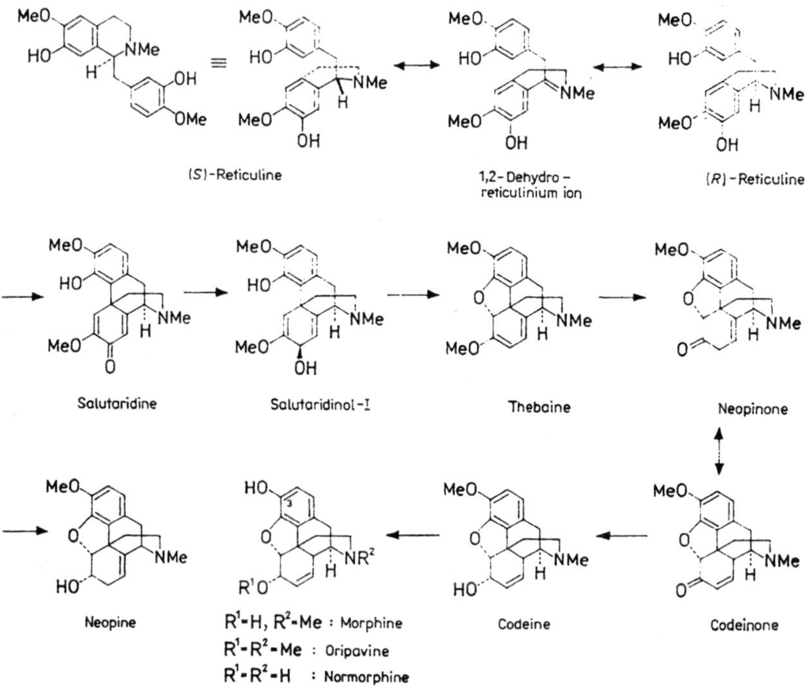

enol ether of thebaine yielding the keto group of neopinone and thence migration of the double bond into conjugation forming codeinone. This is then reduced to codeine.

According to authors[115] codeinone is present to an extent of 5 percent of the amount of thebaine. It appears that the conversion of codeine methyl ether to codeine is an aberrant path, resulting from demethylation by a non-specific demethylating enzyme. The biosynthetic enol ether cleavage of thebaine proceeds by methyl cleavage with retention of the oxygen-6, probably involving an oxygenase. While O-demethylation of enol ether group of thebaine is an enzymatic reaction, the rearrangement of neopinone to codeinone does not require an enzyme. In aqueous solution, an equilibrium is established that favors codeinone. Thus, the O-demethylation of thebaine to morphine occurs at two steps. It is reasonable to hypothesize that these two reactions may be catalysed by different types of enzymes since an enol ether and an aromatic ether are involved in the reactions.

The unnatural thebaine analogue oripavine 3-ethyl ether was efficiently metabolized to morphine 3-ethyl ether and morphine and several unnatural codeine analogues were demethylated to the corresponding morphine analogues in the opium poppy (Figure 9.14).

Trease and Evans[113] further observed that the alkaloids involved in Figure 9.13 are derived from (–)–reticuline. However, (+) reticuline also gives rise to a number of bases: narcotine (noscopine) and narceine of opium; canadine, berberine, and hydrastine of *Hydrastis* (Berberidaceae); and sinomenine (the enantiomer of opium alkaloid salutaridine of *Sinomenium acutum* [Menispermaceae]). With the exception of sinomenine, these alkaloids are termed "berberine bridged" alkaloids and they arise from norlaudanosoline and a one-carbon unit, which is in fact derived from the N-methyl group of (+)-reticuline. The methylenedioxy group of these alkaloids is formed by oxidative cyclization of an O-methoxyphenol function.

BIOSYNTHESIS OF PAPAVERINE

Two alkaloids which arise as branches of the principal biogenetic pathway are neopine, and papaverine which arises by methylations of norrecticuline followed by dehydrogenation. The presence of some of the other minor alkaloids of the opium can be explained by various methylations and dehydrogenations of laudanosoline, reticuline, and their nor derivatives. Various oxidative couplings of reticuline account for other minor alkaloids (e.g., corytuberine and isoboldine).[113]

In 1910 Winterstein and Trier[63] suggested that the benzylisoquinolines were derived from two units of dopa, one of which was transformed into 3,4-dihydroxyphenylacetaldehyde and the second into dopamine. Piclet-Spengler condensation would then afford norlaudanosoline which was also established as a precursor (Figure 9.15).

Battersby and Harper[102] reported when (+) tyrosine labeled was fed to *P. somniferum* only carbons 1 and 3 of papaverine were found to be radioactive–the two units derived from tyrosine being incorporated with nearly equal weight (Figure 9.16).

Battersby et al.[110] observed that labeled N-norlaudanosoline is also a precursor of papaverine in *P. somniferum* since feeding the labeled tetrahydrobenzylisoquinoline base to the plant resulted in 1.5 percent incorporation into papaverine (Figure 9.17).

According to Cordell,[105] the late stages in papaverine biosynthesis were investigated using partially methylated benzyltetrahydroisoquinolines. All four isomers, namely (–) nor-reticuline, (±) nororientaline, (±) norprotosinomenine, and norisoorientaline, were incorporated into papaverine without loss of label, but only nor-reticuline and nororientaline were precursors of tetrahydropapaverine which is itself a good precursor of papaverine. Hence it was concluded that (±) norisoorientaline and (±) norprotosinomenine are not normal precursors but are being incorporated into papaverine by way of an aberrant pathway involving norisocodamine and isopacodine. In confirmation

FIGURE 9.15. A major pathway in the biogenesis of papaverine. Courtesy of the author, W. C. Evans. (*Pharmacognosy*, 12th ed., 1983).

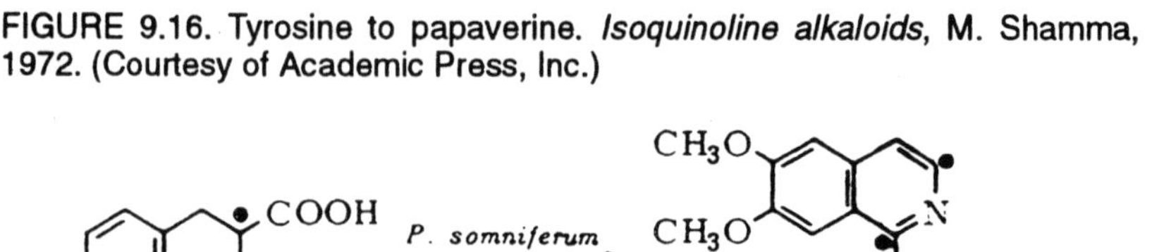

Norreticuline   (–)Norlaudanine   (–)Tetrahydropapaverine   Papaverine

FIGURE 9.16. Tyrosine to papaverine. *Isoquinoline alkaloids*, M. Shamma, 1972. (Courtesy of Academic Press, Inc.)

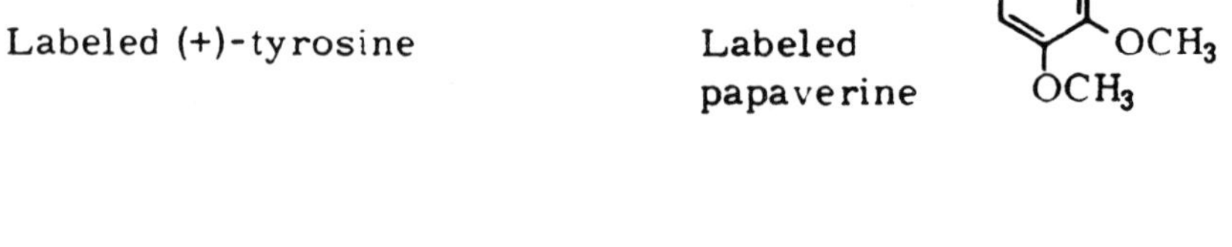

Labeled (+)-tyrosine   Labeled papaverine

FIGURE 9.17. N-norlaudanosoline to papaverine. *Isoquinoline alkaloids.* M. Shamma, 1972. (Courtesy of Academic Press, Inc.)

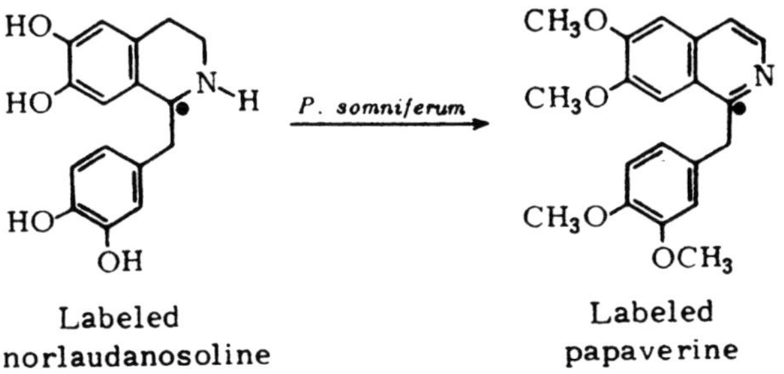

Labeled
N-norlaudanosoline

Labeled
papaverine

of this norprotosinomenine labeled isopacodine to the extent of 12.1 percent. In experiments with (–) and (+) isomers of nor-reticuline and tetrahydropapaverine it was found that only the (–) isomers were precursors of papaverine. The main route to papaverine in *P. somniferum* is therefore from (–) norlaudanosoline via (–) norreticuline or (–) nororientaline to (–) norlaudanidine or (–) norcodamine to (–) tetrahydropapaverine and then to papaverine (Figure 9.18).

Brochmann-Hanssen et al.[104] observed that radioactively labelled (±) nor-reticuline was incorporated into papaverine in *P. somniferum* to a high extent of 5 percent (Figure 9.19).

In their opinion, papaverine arises by methylation of nor-reticuline followed by dehydrogenation to (–) Norlaudanine and then to tetrahydropapaverine which finally leads to papaverine.[99]

BIOSYNTHESIS OF NARCOTINE

Robinson[106] postulated that the phthalideisoquinolines were formed in nature by the oxidative modifications of tetrahydroprotoberberines and the experimental data support this hypothesis. There are about 15 well-characterized phthalideisoquinoline alkaloids isolated from Papaveraceae and Berberidaceae. Narcotine is one of the major alkaloids of opium *Papaver somniferum*, originally isolated by Derosne in 1803.

As observed by Cordell,[105] complementary results were obtained when labeled tyrosines were used as precursors of hydrastine in

FIGURE 9.18. Biosynthesis of papaverine in *P. somniferum.* (G. A. Cordell, 1981). Courtesy of John Wiley & Sons, Inc.

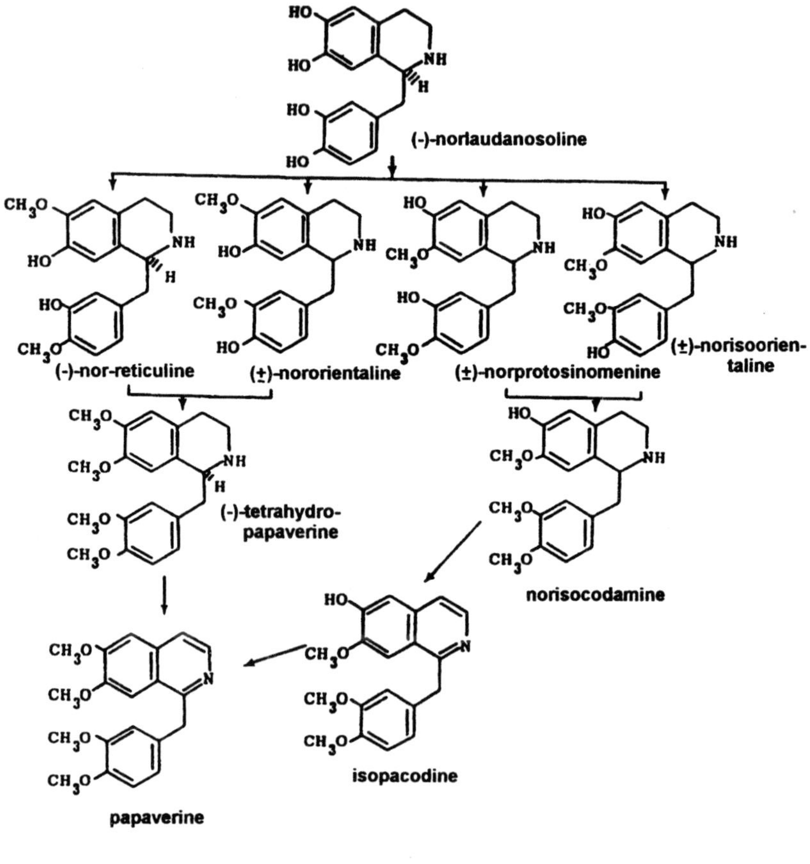

FIGURE 9.19. (±) Norreticuline to papaverine (M. Shamma, 1972). By courtesy of Academic Press, Inc.

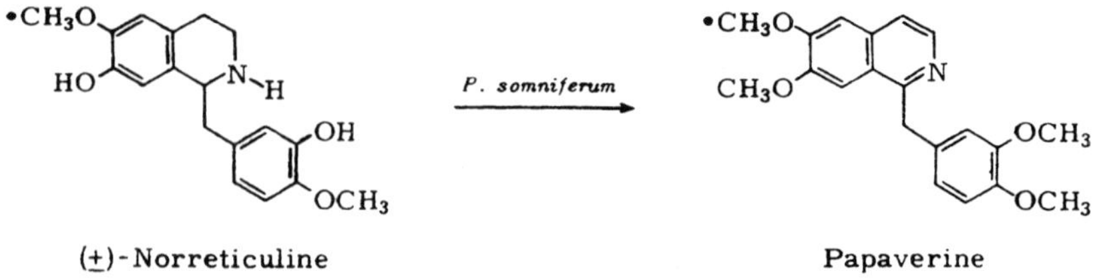

FIGURE 9.20. Biosynthesis of narcotine. (G. A. Cordell, 1981). Courtesy of John Wiley & Sons, Inc.

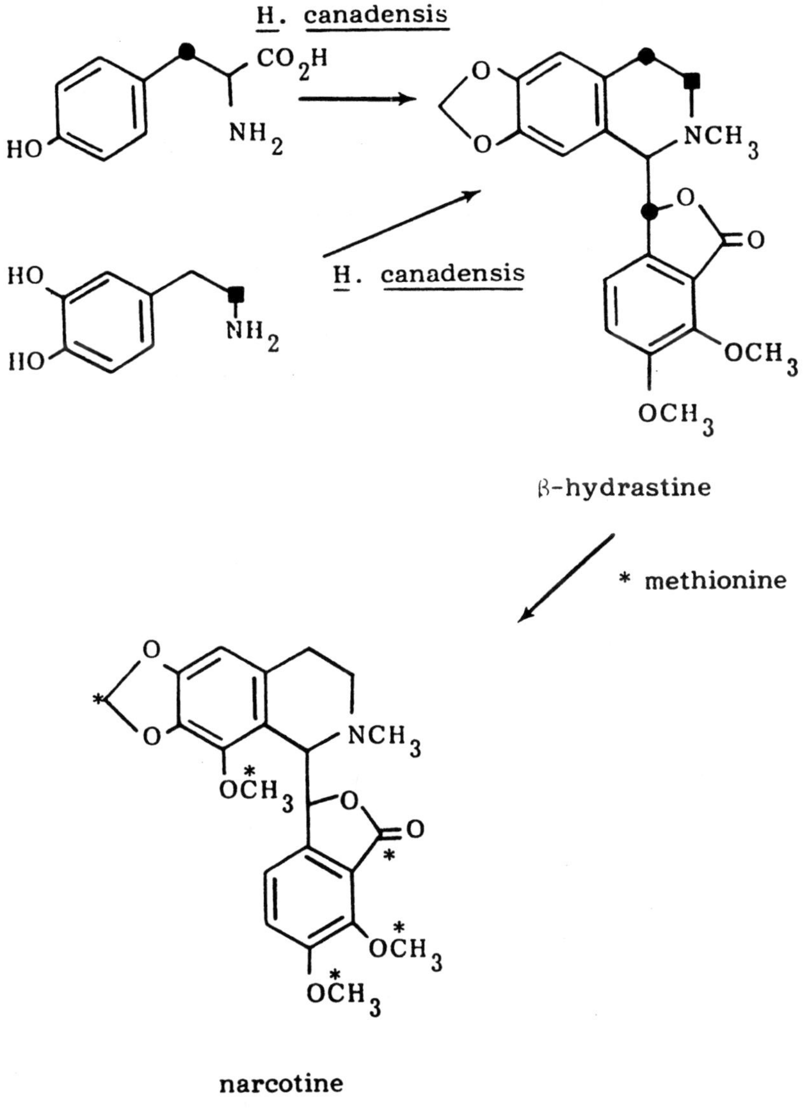

β-hydrastine

narcotine

Hydrastis canadensis and narcotine in *Papaver somniferum*. Thus, [2-[14]] tyrosine labeled C-1 and C-3 of narcotine equally and [3-^{14}C] tyrosine was similarly incorporated into each half of β-hydrastine. [1-^{14}C] Dopamine however labeled only C-3 of the isoquinoline nucleus of hydrastine.[107-109]

Evidence that the carbonyl carbon of narcotine was derived from the berberine bridge carbon came when methionine labeled this carbon as well as the methyl and methylenedioxy groups.[107] Simple benzylisoquinolines such as norlaudanosoline and (+) reticuline were also effective precursors and it is significant to note that label from the N-methyl of reticuline was specifically incorporated into the carbonyl carbon of narcotine.[110] Scoulerine was incorporated to the extent of 4 percent and isocorypalmine 4.6 percent into narcotine.[107,110,111] Canadine was also well incorporated into narcotine suggesting it might be the next intermediate[105] (Figure 9.20).

The step-by-step elucidation of the biogenetic pathway of the opium alkaloids constitutes a brilliant chapter in the history of phytochemical research. Biosynthesis and pathways of some alkaloids have been briefly described in the proceeding pages. The structure elucidation of morphine alkaloids stands as a monument to the brilliance of Robinson but positive proof of the morphine skelton did not come until synthetic endeavors were successfully completed by Gates in 1952.[105]

Brochman-Hanssen[61] rightly concluded, "Looking back over the last 25 years we cannot help but be impressed with what has been accomplished with a combination of chemical mechanistic reasoning, tracer experiments, and isolation work. At the same time, we begin to realize that we have a long way to go before we will understand what transpires at the cellular level. The enzymology of biosynthetic transformation is still in its infancy, but from the growth and development of this branch of plant biochemistry will come major contributions in the future."

REFERENCES

1. Husain, A. and Sharma, J. R. The Opium Poppy. Central Institute of Medicinal and Aromatic Plants, Lucknow, 1983.

2. James, W. O. Alkaloid Formation in Plants, J. Phar. Pharmacol, 5, 1953.

3. Kapoor, L. D. Site of Synthesis of Alkaloids in Some Plants, Curr. Sci., 32, 1963, pp. 355-356.

4. Waller, G. R. and Nowacki, E. K. Alkaloid Biology and Metabolism in Plants. Plenum Press, New York, 1978.

5. Hocking, G. M. A Dictionary of Terms in Pharmacognosy. C. C. Thomas Publisher, Springfield, Illinois, 1955.

6. Clautriau, G. La Localisation des Alcaloides du Pavot. et. de Pharm. et de Chim. (Fr.), 20, 1889, p. 161.

7. Errera, L., Maistriau and Clautriau, G. Localisation et la Signification des Alcaloides dans les Plants, Rec l' Inst. Bot. Bruxelles, 2, 1906, p. 147.

8. Kerbosch, M. G. Bildung und Verbreitung einiger Alkaloide in *Papaver somniferum* L., Arch Pharm. Berlin, 248, 1910, p. 536.

9. Valaer, Peter. 1937, quoted by Fulton, C. C. The Opium Poppy and Other Poppies. U.S. Treasury Dept. Bureau of Narcotics Publications, Washington, 1944.

10. Mika, E. S. Studies on the Growth and Development and Morphine Content of Opium Poppy, Bot. Gaz., 116, 1955, p. 323.

11. Fairbairn, J. W. and Kapoor, L. D. Laticiferous Vessels of *Papaver somniferum* L., Planta medica., 8(1), 1960, p. 49.

12. Kapoor, L. D. Laticiferous vessels of *Papaver somniferum* L. PhD Thesis, London University, 1958.

13. Massicot, J. Biosynthesis of Alkaloids in Poppy Seedlings, Ann. Pharmaceut, France, 19(1), 1961, p. 44.

14. Kovacs, P. and Benesova, M. Some Properties of Aminopetidase from Seedlings of *Papaver somniferum*, Biologia, 31(6), 1960, p. 423.

15. Brastislava, C. Aminopeptidase Activity in Poppy Seedlings, Biochm. Physiol. Pflanzen. 166(2), 1974, p. 113.

16. Maria, B., Senkpiel, K. P. and Barth, A. Multiple Molecular Forms of Aminopeptidases in Poppy Seedlings, Biochem. Physiol. Pflanzen., 175(3), 1980, p. 252.

17. Mothes, K. Physiology of Alkaloids, J. Pharm. Pharmacol., 11, 1959, p. 193.

18. Carpio, M. D. A. "Ensayos sobre la accion c-mitotica de algunos alkaloides del opio en comparacion con la producida por la colchicina." Farmacognosia (Madrid), 7, 1948, p. 83.

19. Dawson, R. F. The Localisation of the Nicotine Synthetic Mechanism in the Tobacco Plant, Science, 94, 1941, p. 396.

20. Crane, F. A. and Fairbairn, J. W. Alkaloids in the Germinating Seedlings of Poppy. Tran. Illinois State Acad. Sci, 63(11), 1970, pp. 86-92.

21. Fairbairn, J. W. and Djote, M. Alkaloid Biosynthesis and Metabolism in an Organelle Fraction in *Papaver somniferum*, Phytochemistry, 9, 1970, pp. 739-742.

22. Kleinschmidt, G. and Mothes, K. Z. Naturforsch, 13B, 1959, p. 52.

23. Kleinschmidt, G. and Mothes, K. Arch. Pharm, 293, 1960, p. 948.

24. Fairbairn, J. W., Hakim, F. and Dickenson, P. B. J. Pharm. Pharmac., 25, 113 p. suppl., 1973.

25. Fairbairn, J. W., Fayha, Hakim and Yahia, El. Kheir Alkaloidal Storage, Metabolism and Translocation in the Vesicles of *Papaver somniferum* Latex, Phytochemistry, 13, 1974, p. 1133.

26. Fairbairn, J. W. Djote, M. and Paterson, A. Biosynthetic Activity of the Isolated Latex, Phytochemistry, 7, 1968, p. 2111.

27. Dickenson, P. B. and Fairbairn, J. W. The Ultrastructure of the Alkaloidal Vesicles of *Papaver somniferum* Latex, Ann. Bot., 39, 1975, pp. 707-712.

28. Fairbairn, J. W. and El-Masry, S. The Alkaloids of *Papaver somniferum* L. VI "Bound" Morphine and Seed Development, Phytochemistry, 7(2), 1968, p. 181.

29. Fairbairn, J. W. Why Do Plants Produce Alkaloids, Pharma International, 4/70, 1979.

30. Pfeifer, S. and Heydenreich, K. On the Metabolism in *Papaver somniferum* L. Hourly Variation in Alkaloid Content, Sci. Pharm. 30, 1962, pp. 164-173.

31. Fairbairn, J. W. and Wassel, G. The Alkaloids of *Papaver somniferum* L. III Biosynthesis in the Isolated Latex, Phytochemistry, 3, 1964, p. 583.

32. Fairbairn, J. W., Paterson, A. and Wassel, G. The Alkaloids of *Papaver somniferum* L. II C^{14} Isotopic Studies of the Rapid Changes in the Major Alkaloids, Phytochemistry. 3(5), 1964, p. 577.

33. Phillipson, J. D., Handa, S. S., and El-Dabbas, S. W. N-oxides of Morphine, Codeine and Thebaine and Their Occurrence in Papaver Species, Phytochemistry, 15, 1976, pp. 129-130.

34. Fairbairn, J. W. and Steele, M. J. Biosynthetic and Metabolic Activities of Some Organelles in Papaver Somniferum Latex. Phytochemistry, 20(5), 1981, pp. 1031-1036.

35. Fairbairn, J. W. and El-Masry, S. S. The Alkaloids *of Papaver somniferum* V Fate of the "End Product" Alkaloid Morphine, Phytochemistry, 6, 1967, p. 499.

36. Fairbairn, J. W., Handa, S. S., Gurkan, E. and Phillipson, J. D. In Vitro Conversion of Morphine to its N-oxide in *Papaver somniferum* Latex, Phytochemistry, 17, 1978, pp. 261-262.

37. Fairbairn, J. W. and Williamson, E. M. Meconic Acid as a Chemotaxonomic Marker in the Papaveraceae, Phytochemistry, 17, 1978, p. 2087.

38. Fairbairn, J. W. and Steele, M. J. Meconic Acid and Alkaloids in *Papaver somniferum* and *P. bracteatum*. Planta Med., 41, 1981, pp. 55-60.

39. Antoun, M. D. and Roberts, M. F. Some Enzymes of General Metabolism in the Latex of *P. somniferum*, Phytochemistry, 14, 1975, pp. 900-914.

40. Antoun, M. D. and Roberts, M. F. Enzymic Studies in *Papaver somniferum*. 5. The Occurence of Methyltransferase Enzyme in Poppy Latex, Planta Med., 28, 1975, p. 6.

41. Jindra, A., Kovacs, P., Pittinerova, Z., Sovova, M. and Sinogrovicova, H. Enzymatic Reactions in Alkaloid Biosynthesis, Herba Hung, 5(2-3), 1966, p. 30.

42. Kovacs, P. and Jindra, A. Biosynthesis of Opium Alkaloids. On the Transformation of Tyrosine to 3.4-dihydroxy-phenylalanine in *Papaver somniferum* Plants, Experientia, 21(1), 1965, p. 18.

43. Wold, J. K., Paulsen, B. S., and Nordal, A. Precursor Incorporation Experiment in Papaver Alkaloid Biosynthesis. II. Acta Pharm. Suec., 14(4), 1977, p. 403.

44. Robinson, Trevor and Nagel, Walter. Peroxidases of *Papaver somniferum*, Phytochemistry, 21, 1982, pp. 535-537.

45. Fairbairn, J. W. and Steele, M. J. Bound Forms of Alkaloids in *Papaver somniferum* and *P. bracteatum*, Phytochemistry, 19, 1980, pp. 2317-2321.

46. Stermitz, F. R. and Rapoport, H. Biosynthesis of Opium Alkaloids. Alkaloid Interconversions in *Papaver somniferum* and *P. orientale*, J. Am. Chem. Soc., 83, 1961, p. 4045.

47. Miller, R. J., Jolles, C. and Rapoport, H. Morphine Metabolism and Normorphine in *Papaver somniferum*, Phytochemistry, 12, 1973, p. 547.

48. Rapoport, H., Stermitz, F. R., and Baker, D. R. The Biosynthesis of Opium alkaloids. I The Interrelationship among Morphine, Codeine and Thebaine, J. Am. Chem. Soc., 82, 1960, p. 2765.

49. Leete, E. Biogenesis of Morphine, J. Am. Chem. Soc. 81, 1959, p. 3948.

50. Martin, R. O., Warren, M. E., and Rapoport, H. The Biosynthesis of Opium Alkaloids. Reticuline as the Benzyltetrahydroisoquinoline Precursor of Thebaine in Biosynthesis with Carbon-14 Dioxide, Biochemistry, 6(8), 1967, p. 2355.

51. Horn, J. S., Paul, A. G., and Rapoport, H. Biosynthetic Conversion of Thebaine to Codeinone. Mechanism of Ketone Formation from Enol Ether *in Vivo*, J. Am. Chem. Soc., 100, 1978, p. 1895.

52. Antoun, M. D. and Roberts, M. F. Methylating and Demethylating Enzymes in *Papaver somniferum* Latex, J. Pharm. Pharmac., 26 suppl., 1974, p. 114.

53. Spencer, I. D. Lloydia., 29, 1966, p. 71.

54. Hodges, C. C. and Rapoport, H. Enzymic Reductions of Codeinone in Vitro, Cell Free Systems from *Papaver somniferum* and *P. bracteatum*, Phytochemistry, 19, 1980, p. 1681.

55. Furuya, T., Nakano, M. and Yoskikawa, T. Biotransformation of (RS) Reticuline and Morphinan Alkaloids by Cell Cultures of *Papaver somniferum*, Phytochemistry, 17, 1978, p. 891.

56. John Tam, W. H., Kurz, W. G. W., Constable, F. and Chatson, K.B. Biotransformation of Thebain by Cell Suspension Cultures of *Papaver somniferum* C. V. Marianne, Phytochemistry, 21(1), 1982, p. 253.

57. Williams, R. D. and Ellis, B. E. Age and Tissue Distribution of Alkaloids in *Papaver somniferum*, Phytochemistry, 28(8), 1989, pp. 2085-2088.

58. Roberts, M. F. Phytochemistry, 10, 1971, p. 3021.

59. Hsu, An-Fei, Liu Ray, H. and Piotrowski, E. G. Conversion of Codeine to Morphine in isolated Capsules of *Papaver somniferum*, Phytochemistry, 24(3), 1985, p. 473.

60. Fairbairn, J. W. and Paterson, A. Alkaloids as Possible Intermediaries in Plant Metabolism, Nature, 5041, 1966, p. 1163.

61. Brochmann-Hanssen, E. Biosynthesis of Morphinan Alkaloids. In The Chemistry and Biology of Isoquinoline Alkaloids, ed. Phillipson et al., © Springer-Verlag Berlin Heidelberg, 1965.

62. Gulland, J. M. and Robinson, R. Constitution of Codeine and Thebaine, Mem. Proc. Manchester Lit, Philos Soc., 69, 1925, pp. 79-86.

63. Winterstein, E. and Trier, G. Die Alkaloide. Borntrager, Berlin, 1910, p. 307.

64. Pummerer, R., Puttfarcken, H. and Schopflocher, P. Oxidation of Phenol. VIII. Dehydrogenation of P. Cresol. Chem. Ber., 5B-B, 1925, pp. 1808-1820.

65. Barton, D. H. R. and Cohen, T. Some Biogenetic Aspects of Phenol Oxidation Festschr Stoll. Birkhauser Basel, 1957, pp. 117-143.

66. Battersby, A. R. and Harper, B. J. T. Rate Studies on the Incorporation of Tyrosine into Morphine, Codeine and Thebaine, Tetrahedron Lett., 27, 1960, pp. 21-24.

67. Gopinath, K. W., Govindachari, T. R., Pai, B. R. and Viswanathan, N. The Structure of Reticuline, a New Alkaloid from *Annona reticulata*, Chem. Ber., 92, 1959, pp. 776-779.

68. Battersby, A. R., Binks, R., Francis, R. J., McCaldin, D. J. and Ramuz, H. Alkaloid Biosynthesis. Part IV. 1-Benzylisoquinolines as Precursors of Thebaine, Codeine, and Morphine, J. Chem. Soc., 1964, pp. 3600-3610.

69. Barton, D. H. R., Kirby, G. W., Steglich, W., Thomas, G. M., Battersby, A. R., Dobson, T. A., and Ramuz, H. Investigations on the Biosynthesis of Morphine Alkaloids, J. Chem. Soc., 1956, pp. 2423-2438.

70. Hodges, C. C. and Rapoport, H. Enzymic Conversion of Reticuline to Salutaridine by Cell Free Systems, Biochemistry, 21, 1982, pp. 3729-3734.

71. Ginsberg, D. C. The Opium Alkaloids. Selected Topics. Interscience, New York, London, 1962, pp. 90-91.

72. Battersby, A. R. Tilden Lecture. The Biosynthesis of Alkaloids, Proc. Chem. Soc., 1963, pp. 189-200.

73. Barton, D. H. R., Bhakuni, D. S., James, R., Kirby, G. W. Phenol Oxidation and Biosynthesis. Part XII. Stereochemical Studies Related to the Biosynthesis of the Morphine Alkaloids, J. Chem. Soc. (c), 1967, pp. 128-132.

73a. Battersby, A. R., Foulkes, D. M., and Binks, R. Alkaloid Biosynthesis Part VIII. Use of Optically Active Precursors for Investigation on the Biosynthesis of Morphine Alkaloids, J. Chem. Soc., 1965, pp. 3323-3332.

74. Battersby, A. R., Evan, G. W., Martin, R. O., Warren, M. E., Jr., Rapoport, H. Configuration of Reticuline in the Opium Poppy. Tetrahedron Lett., 1965, pp. 1275-1278.

75. Borkowski, P. R., Horn, J. S. and Rapoport, H. Role of 1,2-dehydroreticulinium Ion in the Biosynthetic Conversion of Reticuline to Thebaine, J. Am. Chem. Soc., 100, 1978, pp. 276-281.

76. Brochmann-Hanssen, E. and Furuya, T. A New Opium Alkaloid. Isolation and Characterization of (-)-1-(3'-hydroxy-4'-methosybenzyl)-2-methyl-6-methoxy-7-hydroxy-1,2,3,4-tetrahydroisoquinoline [(-)-reticuline], Planta Med., 12, 1964, p. 328.

77. Brochmann-Hanssen, E. and Nielsen, B. (+) Reticuline–A New Opium Alkaloid, Tetrahedron Lett, 1965, pp. 1271-1274.

78. Kamo, K. K. and Mahlberg, P. G. Dopamine Biosynthesis at Different Stages of Plant Development in *Papaver somniferum*, J. Natural Products, Vol. 47, No. 4, 1984, pp. 682-686.

79. Battersby, A. R., Martin, J. A., and Brochmann-Hanssen, E. Alkaloid Biosynthesis Part X. Terminal Steps in the Biosynthesis of the Morphine Alkaloids, J. Chem. Soc. (c), 1967, pp. 1785-1788.

80. Blaschke, G., Parker H. J., and Rapoport, H. Codeinone as the Intermediate in the Biosynthetic Conversion of Thebaine to Codeine, J. Am. Chem. Soc., 89, 1967, pp. 1540-1541.

81. Parker, H. I., Blaschke, G., and Rapoport, H. Biosynthetic Conversion of Thebaine to Codeine, J. Am. Chem. Soc., 94, 1972, pp. 1276-1282.

82. Hohmeyer, A. H. and Shilling, W. L. Isolation and Purification of Neopine, J. Org. Chem., 12, 1947, pp. 356-358.

83. Kirby, G. W., Massey, S. R., and Steinreich, P. Biosynthesis of Unnatural Morphine Derivatives in *Papaver somniferum*, J. Chem. Soc. Perkin Trans., 1, 1972, pp. 1642-1647.

84. Brochmann-Hanssen, E. and Okamoto, Y. Biosynthesis of Opium Alkaloids. Substrate Specificity and Aberrant Biosynthesis. Attempted Detection of Oripavine in *Papaver somniferum*, J. Nat. Prod., 43, 1980, pp. 731-735.

85. Brochmann-Hanssen, E. and Cheng, C. Y. Aberrant Biosynthesis of Opium Alkaloids. Biosynthetic Conversion of the 6-ethyl Analog of Thebaine to Codeine and Morphine, J. Nat. Prod., 45, 1982, pp. 437-439.

86. Nielsen, B., Roe, J. and Brochmann-Hanssen, E. Oripavine–A New Opium Alkaloid, Planta Med., 48, 1983, pp. 205-206.

87. Brochmann-Hanssen, E. A Second Pathway for the Terminal Steps in the Biosynthesis of Morphine, Planta med., 51, 1984, pp. 343-345.

88. Hodges, C. C., Horn, J. S., and Rapoport, H. Morphinan Alkaloids in *Papaver bracteatum*: Biosynthesis and Fate, Phytochemistry, 16, 1977, pp. 1939-1942.

89. Brochmann-Hanssen, E. and Wunderly, S. Biosynthesis of Morphine Alkaloids in *Papaver bracteatum* Lindl., J. Pharm. Sci., 67, 1978, pp. 103-106.

90. Kuppers, F. J. E. M., Salemink, C. A., Bastart, M. and Paris, M. Alkaloids of *Papaver bracteatum*. Presence of Codeine, Neopine and Alpinine, Phytochemistry, 15, 1976, pp. 444-445.

91. Theuns, H. G., Dam, J. E. G. Van, Luteijn, J. M., and Salemink, C. A. Alkaloids of *Papaver bracteatum*: 14-B-hydroxycodeinone, 14-B-hydroxycodein and Methylcorydaldine, Phytochemistry, 16, 1977, pp. 753-755.

92. Meshulam, H. and Lavie, D. The Alkaloidal Constituents of *Papaver bracteatum* Arya II, Phytochemistry, 19, 1980, pp. 2633-2635.

93. Miram, R. and Pfeifer, S. On the Changs in the Alkaloid Content of Opium Poppy during a Growing Season, Sci. Pharm., 28, 1960, pp. 15-28.

94. Fairbairn, J. W. and Wassel, G. The Alkaloids of *Papaver somniferum* L. I Evidence for a Rapid Turnover of the Major Alkaloids, Phytochemistry 3, 1964, pp. 253-258.

95. Neubauer, D. Distribution of the Major Alkaloids of the Opium Poppy in the Various Parts of the Plant at Different Stages of Development, Planta med., 12, 1964, pp. 43-50.

96. Wold, J. K. Bound Morphine and Codeine in the Capsule of *Papaver somniferum*, Phytochemistry, 17, 1978, pp. 832-833.

97. Brochmann-Hanssen, E. and Cheng, C. Y. Biosynthesis of a Narcotic Antagonist. Conversion of N-allylnorreticuline to N-allylnormorphine in *Papaver somniferum*, J. Nat. Prod., 43, 1984, pp. 731-735.

98. Brochmann-Hanssen, E., Cheng, C. Y. and Chiang, H. C. Biosynthesis of Opium Alkaloids. The Effects of Structural Modifications of Reticuline on Racemization and Biotransformation, J. Nat. Prod., 45, 1982, pp. 629-634.

99. Brochmann-Hanssen, E., Chen, C. H., Chen, C. R., Chiang, H. C. Opium Alkaloids, Part XVI The Biosynthesis of 1-benzylisoquinolines in *Papaver somniferum*. Preferred and Secondary Pathways, Stereochemical Aspects, J. Chem. Soc. Perkin Trans., 1, 1975, pp. 1931-1937.

100. Roberts, M. F. and Antoun, M. D. Phytochemistry, 17, 1978, p. 1083.

101. Shamma, Maurice. The Isoquinoline Alkaloids. Academic Press, New York, 1972.

102. Battersby, A. R. and Harper, B. J. T. J. Chem. Soc. London, 1962, p. 3526.

103. Battersby, A. R., Binks, R., Francis R. J., McCaldin, D. J. and Ramuz, H. J. Chem. Soc. London, 1964, p. 3600.

104. Brochmann-Hanssen, E., Fu, C. C., Leung, A. Y. and Zanati, G. J. Pharm. Sc., 60, 1971, p. 1672.

105. Cordell, Geoffrey, A. Introduction to Alkaloids. John Wiley & Sons, New York, 1981.

106. Robinson, R. The Structural Relations of Natural Products, Oxford University Press (Clarendon) London & New York, 1955.

107. Battersby, A. R., Hirst, M., McCaldin, D. J., Southgate, R. and Stauton, J. J. Chem. Soc., C. 1968, p. 2163.

108. Gear, J. R. and Spencer, I. D., Biochem. Biophys. Res. Commun., 13, 1963, p. 115.

109. Battersby, A. R. and Hirst, M. Tetrahedron Lett., 1965, p. 669.

110. Battersby, A. R., Francis, Hirst, M., Southgate, R. and Staunton, J. Chem. Commun., 1967, p. 602.

111. Parker, H., Blaschke, G. and Rappoport, H. J. Am. Chem. Soc., 94, 1972, p. 1286.

112. Hsu, An-Fei and Pack, Judith. Metabolism of ^{14}C-codein in Cell Cultures of *Papaver somniferum*, Phytochemistry, 28(7), pp. 1879-1881.

113. Trease, G. E. and Evans, W. C. Pharmacognosy, 12th Edition. Bailliere Tindal London, 1983.

114. Swain, G. E. An Introduction to Alkaloids. John Wiley & Sons, Inc., New York, 1967.

115. Schutte, H. R. and Liebisch, H. W. Alkaloids Derived from Tyrosine and Phenylalanine in Biochemistry of Alkaloids, ed. K. Mothes, H. R. Schutte and M. Luckner. VEB Deutscher Verlag der Wissenschaften Berlin/DDR, 1985.

116. Griffing, L. R. and Nessler, C. L. Immunolocalization of the Major Latex Proteins in Developing Laticifers of Opium Poppy (*Papaver somniferum*), J. Plant Physiol., 134(3), 1989, pp. 357-363.

Chapter 10

Occurrence and Role of Alkaloids in Plants

The history of alkaloid research begins with morphine, a major alkaloid of opium poppy (*Papaver somniferum*). Since the isolation of morphine in the early nineteenth century there has been a vigorous search for alkaloids or "plant alkalies" from the vegetable kingdom. After more than ten years Sertürner's discovery led to the isolation of a series of further basic substances from other physiologically active natural products. As a result such alkaloids as strychnine, emetine, brucine, piperine, caffeine, quinine, cinchonine, and colchicine were obtained during 1817-1820 in the laboratory of Pelletier and Caventou at the faculty of Pharmacy in Paris,[1] in the following chronological order:[2]

Narcotine	– 1817 by Robiquet
Emetine	– 1817 by Pelletier and Magendie
Strychnine	– 1818 by Pelletier and Caventou
Veratrine	– 1818 by Meibner
Brucine	– 1819 by Caventou and Pelletier
Piperine	– 1819 by Oersted
Caffeine	– 1820 by Runge
Cinchonine and Quinine	– 1820 by Pelletier and Caventou

But the pace of alkaloid research did not pick up until early 1900. In 1900 A.D. Pictet[42] reported about 100 alkaloids whose chemical structures were all more or less known. In 1908 Euler[43] described about 100 alkaloids including about 30 with unknown chemical structure. In 1922 Wolffenstein[44] reported about 150 alkaloids at least partially known and 100 chemically unknown or almost unknown alkaloids. By 1939 nearly 300 alkaloids had been isolated and about 200 of these had at least reasonably well-defined structures.[1] With the introduction of preparative chromotographic techniques and sophisticated spectroscopic instrumentation, the number of the known alkaloids has risen dramatically. In 1960 Boit[45] described about 2,000 alkaloids including about 600 with unknown chemical structure. Raffauf[46] listed about 4,000 alkaloids in 1970. A review of the middle of 1973 counted 4,959 alkaloids, of which 3,293 had known structures.[1] Waller and Nowaki[3] estimated in 1978 the number of known alkaloids to be about 6,000,800 of which had been discovered in the two preceding decades. In 1985, Mothes et al.[2] observed about 7,000 more or less well characterized alkaloids may have been known. This number may have swelled since then.

Verpoorte et al.[55] observed that plants have a variety of alkaloids. Table 10.1 presents some of the major classes and an estimation of the number of representatives.

Isolation of alkaloid is incomplete without structure determination and this final step was generally the synthesis of the alkaloid in question and successful synthesis were important milestones in the history of alkaloid research. For instance, morphine isolated in 1806 was synthesized in 1952 by Gates and Tschudi.[4]

OCCURRENCE

It is true that alkaloids are widely distributed throughout the plant kingdom, but no facile generalization about the alkaloid distribution can be made.[54] There is some tendency for higher plants to have more alkaloids than lower plants, but alkaloids are well known in the club mosses (*Lycopoduim spp.*) and horsetails (*Equisetum spp.*) and certain fungi (ergot).

Robinson[5] observed that distribution of alkaloids in nature, while not completely random, cannot be described in any simple and unambiguous way. Higher plants are the chief source of alkaloids. Surveys of ferns and bryophytes have revealed that alkaloids are lacking or extremely rare.

Among the seed plants a greater variety of alkaloids has been

TABLE 10.1. Number of unique structures in major alkaloid classes in NAPRAL-ERT database.[a]

Class.	Number
INDOLE	4125
Isoquinoline	4045
Protoalkaloids	1147
Quinoline	718
Pyrrolidine/Piperidine	714
Diterpene	642
Quinolizidine	571
Pyrrolizidine	562
Steroidal	440
Tropane	300
Pyridine	241
Indolizidine	170
Sesquiterpene	132
Homospermidine/spermidine	129

a. October 1988. Totals include alkaloids isolated from plants, animal, and marine sources.

found in Dicotyledons than in Monocotyledons or gynosperms. It has been estimated that about 15-30 percent of all plants species contain alkaloids.[47-53] Not a single alkaloid has been found among the thousand of plants in the orders Pandanales, Salicales, and Fagales. In the family Solanaceae about two-thirds of the species have alkaloids. A famous alkaloidal family, the Apocynaceae, has produced about 900 different alkaloids.

Alkaloid-producing plant families are found in the tropics, about three times more than in temperate zones, but this reflects the greater number of tropical plants generally since the ratio of tropical alkaloid families to total tropical families is the same as the ratio for temperate.

A statistical analysis of 3,600 alkaloid plants showed caffeine occurring in the largest number of families, lycorine in the largest number of genera, and berberine in the largest number of species.[6]

Cordell[1] reported that out of 60 orders of the Engler plant system,[34] contain alkaloid-bearing species. Forty percent of all plant families contain at least one alkaloid-bearing plant. However, alkaloids are reported in only 8.7 percent of over 10,000 genera. The most important alkaloid-bearing families are the Liliaceae, Amaryllidaceae, Compositae, Ranunculaceae, Menispermaceae, Lauraceae Papaveraceae, Leguminosae, Rutaceae, Loganiaceae, Apocynaceae, Solanaceae, and Rubiaceae. He observed that in most plant families that contain alkaloids, some genera contain alkaloids whereas others do not. The Papaveraceae is an unusual family in that all the species of all the genera so far studied contain alkaloids.

A given genus will often yield the same or structurally related alkaloids, and even several different genera within a family may contain the same alkaloid. For example, hyoscyamine has been obtained from seven different genera of the plant family Solanaceae. In addition, simple alkaloids such as nicotine frequently occur in botanically unrelated plants. On the other hand the more complex alkaloids, such as vindoline and morphine, are often limited to one species or genus of the plant.

Robinson[9] observed that an alkaloidal plant can be defined as one that has more than 0.05 percent alkaloid by dry weight. By processing prodigious amounts of material it has been found that the "non-alkaloidal" common cabbage contains 0.0004 percent narcotine,[47] and by radio immunoassay it appears that morphine may be widely distributed in plants at levels less than 1ng/g fresh weight.[48] The highest concentration of alkaloid ever found was in a strand of *Senecio riddelii* that had content of 18 percent.[49]

Another kind of arbitrariness arises from the fact that even well-established alkaloid plants may not contain alkaloids in all tissues or at all stages of development and in all geographical locations.[50,51,52]

Major source of alkaloids in the past has been the angiosperms.[1] However, in recent years there have been increasingly numerous examples of the occurrence of alkaloids in animals, insects, marine organisms, microorganisms, and the lower plants. At least 50 alkaloids have been isolated from animal organs but only 12 of these are characteristic of animals alone, the remainder are common to those of plant origin.[7] Two hundred and fifty-six alkaloids are distributed throughout the lower plant orders as follows: Agaricaceae (34), Hypocreaceae (51), other fungi (40), Algae (5), Cycadaceae (6), Equisetaceae (5), Lycopodiaceae (95), Pinaceae (2), Gnetaceae (9), Taxaceae (9). A number of simple nitrogenous principles found in algae are not recognized as true alkaloids by some authorities.

Some examples of this very diverse occurrence of alkaloids are the isolation of muscopyridine from the musk deer; castoramine from the Canadian beaver; the pyrrole derivative, a sex pheromone of several insects; saxitoxin, the neurotoxic constituent of the red tide *Gonyaulax catenella*; pyocyanine from the bacterium *Pseudomonas aeruginosa*, chanoclavine-1 from the ergot fungus *Claviceps purpurea,* and lycopodine from the genus of club mosses *Lycopodium.*[1]

FUNCTIONS

Many of these alkaloids have dramatic effects on animal and human bodies in marked contrast to their apparently passive role in the plant.[8] It has not been explained satisfactorily why such chemicals as strychnine, atropine, d-tubocurarne, morphine, or cocaine should be so quiescent in the plant but immediately on transfer to animals quite striking effects take place. It has been suggested that animals, unlike plants, possess sensitive nervous systems which react in various ways to the alkaloids. These responses have been of great use in therapeutics for alleviating the cause of human suffering.

The question "Why are alkaloids found in plants?" remains unanswered. Do they serve any function at all, apart from man's use of them in therapy? The function of alkaloids in plants is obscure although many suggestions have been made.[9]

One of the earliest suggestions was that alkaloids function as nitrogen waste products like urea and uric acid in animals. This idea of a waste product of nitrogen metabolism "flotsam thrown up on the metabolic beach" as Tschirch put it, was dismissed.[8] Some have suggested a protective function as the presence of alkaloids would discourage animals from eating the plants. Mothes[10] has shown that alkaloidal plants are more numerous in regions of high grazing by animals, such as in desert oasis or alpine clearings. On the other hand, it is well known that rabbits, hares, etc., eat belladonna plants; opium poppy plants are regularly eaten by rabbits in the spring. Flocks of goats are seen in Sudan feeding, with evident relish, on the poisonous *Datura metal* and *Argemone mexicana* growing on the banks of the Nile. If, however, some alkaloids do impart "survival value" to plants, it is almost certainly due to their bitterness rather than to their toxicity.[8] Although evidence favoring protective function has been brought forward in some instances, it is probably an overworked and anthropocentric concept.[9] Many alkaloids poisonous to man have no effect on other (and more significant) enemies of the plants. An example is bulbous agaric which contains highly poisonous peptides that cause the death of many human beings every year. Roe deer, however, may feed on this mushroom without being affected.[2] Another example is of pyrrolizidine alkaloids formed in plants, which are poisonous to most animals but are tolerated by some specialized insects that may even metabolize these substances in the production of pheromones.

Some workers have suggested that alkaloids act as nitrogen reserve or even as a method of detoxification of harmful nitrogen metabolites. Some alkaloids may serve as nitrogen storage reservoirs although many seem to accumulate and are not further metabolized even in severe nitrogen starvation.[9]

Robinson[5] observed that the role of alkaloids as detoxication products appears plausible in some instances, as in the case of alkaloids that may remove such active molecules as indole-3-acetic acid or nicotinic acid from sites where they could unbalance metabolism. Because of great diversity among alkaloids it seems likely that theories which give plausible functions to several of them cannot be universally applicable.

Both Mothes[10] and James,[11] in reviewing their own work, agreed that alkaloids can scarcely be called waste products but neither can they be given a definite function at the present. Their guess was that they were sort of halfway up the evolutionary scale: on their way to being involved into molecules of significance to the plant.

Another suggestion is that alkaloids may serve as growth regulators since structures of some of them resemble structures of known growth regulators. Certain plant hormones may act by virtue of their chelating ability and some alkaloids may also possess chelating ability.[9]

It has been shown that lupin alkaloids may act as germination inhibitors.[12] On the other hand, Jacquiot[13] has shown that alkaloids may remove the inhibitory effects of tannins on growth of plant tissue cultures. Alkaloids do have effect on plant growth-usually inhibiting such processes as elongation or seed germination.[33,12,34] Some of these inhibitions could be significant in the competition between plants.[35]

It was originally suggested that alkaloids, being mostly basic, might serve in the plant to replace mineral bases in maintaining ionic balance. Dawson[14] observed that feeding nicotine to tobacco root cultures increased their uptake of nitrate. Another view is that alkaloids function by exchanging with soil cations and it has been found that alkaloids are excreted by the roots of several alkaloid plants.[15]

Waller and Nowaki[3] pointed out the metabolic status of alkaloids. They repeat the most common notion that alkaloids are waste products which play an unimportant role as plant protective compounds and accumulate in the plant because the plant lacks an effective excretory organ comparable to the animal kidney. In their opinion the plants do have a surprisingly effective system–the aging leaf. Curiously enough this system is not widely recognized. Each plant, without exception, produces new leaves and drops old ones while growing. Before the leaf is ready to drop, all metabolites which can be useful are translocated and only true waste remains. If alkaloids were waste, they would be concentrated in the dead leaves or be converted into another form as what happens with introduced foreign alkaloids and unusual biosynthetic compounds. It seems more convincing that, since the alkaloid character is well established in some plants, it is necessary for those species to produce and save some alkaloids in, for example, the seeds, as well as traces in other parts.

An inefficient system would long since have been eradicated by selection. The question of the metabolic status of alkaloids arises because only a few of the plants accumulate alkaloids. There is no adequate answer to this question. The role of singular alkaloids may be different in different plants, and in alkaloid-free plants it may be fulfilled by other compounds such as unusual amino acids, other metabolites, etc., but the protective function, both present and past, may be the *raison d'etre* for most alkaloids.

Fairbairn[8] observed that alkaloids are not the "end-products" or slowly accumulating heaps of metabolic sludge but are themselves intermediates in the formation of other substances. He advocates that there should be increased emphasis on the "post alkaloidal" status. Post alkaloidal substances may act as co-enzyme moieties in Conium and may have important functions in seed viability in *Papaver.* Pfeifer and Heydenreich[16] have reported fluctuations in the alkaloidal pattern in *Papaver somniferum* and Tso and Jeffrey[17] have shown that radioactive nicotine when fed to tobacco plants is converted into radioactive amino acids, sugars, and other metabolites.

The major alkaloids of opium poppy, viz, morphine, codeine, thebaine, papaverine, and narcotine, varied rapidly at short intervals and diurnal pattern being superimposed on the variations in the developing capsule. Biosynthesis studies showed that thebaine is formed into codeine and codeine into morphine irreversibly. The major alkaloid morphine did "disappear" at intervals.

It was assumed that soon after its formation from thebaine and codeine, morphine is rapidly metabolized into new substances, some alkaloid-like and other non-alkaloidal "polar" molecules. Some of these bound forms are translocated to the developing ovules and later stored in the seeds as bound alkaloids. During germination they evidently break down again.

It was shown that isolated latex of opium poppy will biosynthesize morphine from dihydroxyphenylalanine *in vitro*. It was also shown that the latex contains viable mitochondria and a heavier fraction containing larger organelles. When the latex is centrifuged at about 3000g these heavy particles form a pellet at the bottom of the tube; the supernatant fluid contains the mitochondria, ribosomes, enzymes, and other smaller molecules but quite surprisingly no alkaloids are present. These are always associated with the heavy organelle fraction. This heavy fraction, when separated from the rest of the latex, will synthesize morphine from radioactive dihydroxyphenylalanine. It is suggested that the entire synthesis of the opium alkaloids from simple amino acids is carried out by specially designed organelles in this plant.[8] The alkaloidal vesicles in the laticifers appear to form by a dilation in the endoplasmic reticulum.[20]

If the plant uses energy and important metabolites and possibly special organelles to produce alkaloids surely some use will be made of them.

It has been shown that the bound forms of the alkaloids are essential to the early stages of seed development, later, when the leaves open out the plant begins to produce its own alkaloids and then can develop normally.

Schutte and Liebisch[21] have reported that morphine shows a rapid turnover in the opium poppy.[22,16] Normorphine is an active metabolite of morphine in *P. somniferum*.[23] Feeding labeled normorphine showed that the N-demethylation step is irreversible. Probably the morphine degradation pathway involves an initial demethylation to normorphine, which is subsequently degraded to non-alkaloidal metabolites. The inclusion of normorphine as the final step in the alkaloid biosynthesis completes the sequence of demethylation from thebaine. It is possible that only the stem latex is involved with the morphine and normorphine activity.[24] Indeed it was found that senescent, dying plants had almost no normorphine despite the fact that morphine concentration was still quite high.

The alkaloid contents increase rapidly at the time of cell enlargement and vacuolization and the increase is followed by a slow decline in concentration during senescence. The initiation of flowering may stop or inhibit alkaloid formation[25] or stimulate it.[26] A young leaf on an old plant may reflect in its alkaloid content the plant's stage of development rather than its own[27] but in some perennial woody plants leaves follow a pattern of declining concentration through the growing season[28,29]– even to containing no detectable alkaloid in the fall.[28] There is no consistent pattern for the ontogeny of alkaloids in developing seeds and germinating seedlings. In some species the alkaloid concentration may decrease after fertilization, giving mature seeds with little or no free alkaloid (e.g., *Nicotiana, Papaver, Hordeum, Datura, and Erythroxylum* spp.). Some seeds contain bound alkaloids (e.g., *Papaver somniferum*).[30,31]

During germination alkaloid synthesis may begin within a few days (*Hordeum*) or only after several weeks (*Datura*). There have also been observed cyclic diurnal variations in alkaloid content, with certain alkaloids more than doubling or halving their concentration within a few hours.[32]

It is evident that biosynthesis of alkaloid molecules must require energy and in some instances the presence of highly specific enzymes. The perpetuation of such a low entropy system through the course of evolution seems to call for an explanation in terms of useful function, but no generally adequate explanation has been forthcoming.[5]

It has been suggested that the formation of volatile terpines somehow competes with the formation of alkaloids so that plants having one lack the other.[36] Tallent and Horning[37] found that species of pine which have alkaloids also have straight-chain aliphatic hydrocarbons rather than terpenes in their terpentines.

As it has not been explained satisfactorily why the production of carbohydrates, fatty acids, aromatic oils, gums and resins, rubber and waxes, etc., is confined to specific plants in specific genera of specific families, so is the case with the formation of alkaloids.

It has been shown that remarkably rapid changes in the pattern and concentration of alkaloids occur during the development of fruit of *Conium maculatum* and *Papaver somniferum*. These would seem to be unlikely to occur if the alkaloids are slowly accumulating waste products. It has also been shown that if radioactive tyrosine is introduced into poppy latex, the alkaloids become radioactive.[38]

On the other hand, since quite early in the present century much speculation has been done on *how* plants produce alkaloids. Robinson[39] was a pioneer and has put forward many important ideas on alkaloid biogenesis. These speculations were based largely upon structural similarities within the alkaloid series and also upon the relations of alkaloids to simpler natural products. They constitute a comprehensive and reasonable picture of alkaloid biosynthesis but are based primarily on analogy with reactions of organic chemistry and by consideration of structural similarity rather than direct, biochemical evidence.[9] The assumption of Robinson's scheme of biosynthesis is that the basic skeletons of alkaloids are derived from common amino acids and other small biological molecules. A few simple types of reactions suffice to form complex structures from these starting materials. Although such approaches proposed by Robinson were not claimed to prove that plants necessarily followed the suggested biosynthetic schemes, nevertheless, these proposals have given help to structural studies on new alkaloids and led to experiments on living plants. They have also led to many new laboratory syntheses of alkaloids under mild so-called "physiological" conditions. Little experimental proof regarding the biosynthetic pathways by which alkaloids are formed was forthcoming before the use of radioactive tracers. Since around 1951, however, an ever-increasing flow of papers on the subject has occurred.[38] Robinson[5] observed that the crucial problem in alkaloid biosynthesis is the identification of the point in the pathways of metabolism where an intermediate is formed whose subsequent transformations are directed solely to an alkaloid with no links to other classes of compounds.

The types of compounds suggested by Robinson[39] do occur in plants, and the required reactions take place so readily that large structures similar to alkaloid skeletons may be formed *in vitro* by mere mixing of the supposed precursors. On the other hand, Wenkert[40,41] believes that alkaloid formation may be more closely related to carbohydrate and acetate metabolism than to amino acid metabolism.

We have still to find a convincing explanation as to why alkaloids are produced predominately in Apocynaceae or Solanacaea, etc., why essential oils are produced by Labiatae and Rutaceae, why edible oils are restricted to Cruciferae, papayeraceae, and why cereals are mainly produced by Gramineae.

Similarly, there is a long list of such compounds, viz., sugar, waxes, rubber, dyes, fiber (cotton and jute), etc., which are produced by specific plants of specific families. Is there any correlation, chemotaxanomic or genetic, or is this just a selection at random by nature?

REFERENCES

1. Cordell, G. A. Introduction to Alkaloids. John Wiley and Sons, New York, 1981.

2. Mothes, K., Schutte, H. R. and Luckner, M. Biochemistry of Alkaloids. VEB Deutscher Verlag der Wissenchaften, Berlin/DDR, 1985.

3. Waller, G. R. and Nowacki, E. K. Alkaloid Biology and Metabolism in Plants. Plenum Press, New York, 1978.

4. Gates, M. and Tschudi, G. J. The synthesis of morphine. Am. Chem. Soc., 74, 1952, p. 1109.

5. Robinson, T. The Biochemistry of Alkaloids, 2nd Edition. Springer-Verlag Berlin, Heidelberg, New York, 1981.

6. Willaman, J. J. and Li, H. L. General relationships among plants and their alkaloids. Econ. Bot., 17, 1963, pp. 180-185.

7. Raffauf, R. F. Some Notes on the Distribution of Alkaloids in the Plant Kingdom, Economic Bot., 24, 1970, pp. 34-38.

8. Fairbairn, J. W. Why Do Plants Produce Alkaloids. Pharma international. Special print of number 4/70. Verlag G. Braun, 75 Karlsruhe 1, Postfach 1709.

9. Robinson, Trevor. The Organic Constituents of Higher Plants. Sixth edition. Cardus Press, North Amherst, 1991.

10. Mothes, K. Ann. Rev. Plant Physiology, 6, 1955, p. 393.

11. James, W. O. J. Pharm. Pharmacol., 5, 1953, p. 809, Endeavor, 12, 1953, p. 76.

12. Maisuryan, N. A. Doklady Akad. Nauk Armyan S.S.R., 22, 1956, p. 91 (Chem. abstr. 50, 10836).

13. Jacquiot, C. Comp. Rend., 225, 1947, p. 434.

14. Dawson, R. F. Plant Physiol., 21, 1946, p. 115.

15. Laroze, A. Alves da Silva, J. Anais. Fac. Farm Porto., 12, 1952, p. 85 (Chem. Abstr. 48, 233).

16. Pfeifer, S. and Heydenreich, K. Sci. Pharm. 30, 1962, p. 164.

17. Tso, T. C. and Jeffrey, R. N. Arch. Biochem. Biophys., 92, 1961, p. 253.

18. Kleinschmidt, G. and Mothes, K. Arch. Pharm., 293, 1960, pp. 948.

19. Meissner, L. and Mothes, K. Phytochem., 3, 1964, p. 1.

20. Nessler, C. L. and Mahlberg, P. G. Am. J. Bot., 64, 1977, pp. 541-551.

21. Schutte, H. R. and Liebisch, H. W. Alkaloids Derived from Tyrosine and Phenylalanine, in Biochemistry of Alkaloids, ed. Mothes, K., Schutte, H. R. and Luckner, M. VEB Deutscher Verlag der Wissenschaften Berlin/DDR, 1985.

22. Fairbairn, J. W. and El Masry, S. Phytochemistry, 6, 1967, p. 499.

23. Colomas, J., Layton, J. L. and Bulard, C. C. R. Acad. Sci. Ser. D277, 1973, p. 173.

24. Fairbairn, J. W. and Djote, M. Phytochemistry, 9, 1970, p. 739.

25. Mothes, K. Alkaloids in Plants, in The Alkaloids Vol. VI, pp. 1-29, R. H. F. Manske and H. L. Holmes, ed. Academic Press, New York, 1960.

26. Muraveva, D. A. and Figurkin, B. A. Biol. Nauki, 15, 1972, pp. 87-89.

27. James, W. O. Alkaloids in Plants, in The Alkaloids, R. H. F. Manske and H. L. Holmes Ed., Vol. 1, pp. 71. Academic Press, New York, 1950.

28. Ripperger, H. Jasminidin, Em Neues Monsterpenalkaloid Aus Syringa Vulgaris Phytochemistry, 17, 1978, pp. 1069-1070.

29. Ziyaev, R., Abdusamatov, A. and Yunusov, S. Y. Chem. Nat. Compds., 11, 1975, pp. 478-481.

30. Grove, M. D., Spencer, G. F., Wakeman, M. V. and Tookey, H. L. J. Agri. Food. Chem., 1976, pp. 896-897.

31. Weeks, W. W. and Bush, L. P. Plant Physiol., 53, 1974, pp. 73-75.

32. Robinson, T. Science, 184, 1974, pp. 430-435.

33. Evenari, M. Bot. Rev., 15, 1949, pp. 153-194.

34. Ramshorn, K. Flora, (Jena), 142, 1955, pp. 601-618. (Chem. Abstr. Plant physiological effects of alkaloids. Influence of nicotine on the elongation of Avena coleoptiles. 52, 1958, pp. 1475-1479).

35. Overland, L. Am. J. Bot., 53, 1966, pp. 423-432.

36. Treibs, W., Sitzber, deut. Akad. Wiss. Berlin, Kl. Math. u. allgem. Naturw, 1953.

37. Tallent, W. H. Horning, E. C. J. Am. Chem. Soc., 78, 1956, p. 4467.

38. Swain, G. R. An Introduction to Alkaloids. John Wiley and Sons, Inc., New York, 1967.

39. Robinson, Robert. The Structural Relations of Natural Products. Clarendon Press, Oxford, 1955.

40. Wenkert, E. Experientia, 15, 1959, p. 165.

41. Wenkert, E. J. Am. Chem. Soc., 84, 1962, p. 98.

42. Pictet, A. Die Pflanzenalkaloide und ihre chemische Konstitution. 2d ed. Springer-Verlag, Berlin, 1900.

43. Euler, H. Grundlagen und Ergebnisse der Pflanzenchemie. Part 1. Friedr. Vieweg & Sohn, Braunschweig, 1908.

44. Wolffenstein, R. Die Pflanzenalkaloide und ihre chemische Konstitution. 3d ed. Springer-Verlag, Berlin, 1922.

45. Boit, H. H. Ergebnisse der Alkaloidchemie bis 1960. Akademie-Verlag, Berlin, 1961.

46. Raffauf, R. F. A Handbook of Alkaloids and Alkaloid Containing Plants. Wiley-Inter-Science, New York, 1970.

47. Hegnauer, R. in T. Swain, ed. Chemical Plant Taxonomy. Academic Press, New York, 1973, pp. 389-427.

48. Hazum, E. and five others. Science, 213, 1981, pp. 1010-1012.

49. Molyneux, R. J. and Johnson, A. E. J. Nat. Prod.,47, 1984, pp. 1030-1032.

50. Oksman-Caldentey, K. M., Vuorela, H., Strauss, A. and Hiltunen, R. Planta Med., 53, 1987, pp. 349-354.

51. Lievy, A., Milo, J. and Palevitch, D. Planta Med., 54, 1988, pp. 299-301.

52. Johnson, A. E., Molyneux, R. J. and Merill, G. B. J. Agric. Food Chem., 33, 1985, pp. 50-55.

53. Smolenski, S. J., Silinis, H., and Farnsworth, N. R. Lloydia, 38, 1975, pp. 411-441.

54. Hegnauer, R. Phytochemistry, 27, 1988, pp. 2423-2427.

55. Verpoorte, Robert, Van der Heejden, Robert, Van Gulk, Walter M., and Ten Hoopen, Hens J. G. Plant Biotechnology for production of alkaloids present status and prospects. In The Alkaloids, Arnold Brossi, ed. Vol. 40, 1991, pp. 1-187. Academic Press Inc., San Diego, California.

Chapter 11

Evaluation of Analgesic Actions of Morphine in Various Pain Models in Experimental Animals

Krishnaswami Ramabadran
Mylarrao Bansinath

INTRODUCTION

Pain is considered as the perfect misery and therefore the control of pain in human subjects has always been a challenging problem for both humanitarian and scientific reasons. As a consequence, extensive basic research has been carried out on laboratory animals. However, these efforts have yielded few consistently effective tools for the physician.[1] Despite the high scientific interest in pain mechanisms, research efforts have so far failed to achieve a level of knowledge that can be considered as a beginning of the approach to our understanding, unlike in other modality of senses (olfaction, vision, hearing, and gustation). Thus efforts to devise and discover new and effective treatments of pain have generally failed. This review addresses some of the important aspects of different *in vivo* methods currently used in experimental animals to study pain. One general issue of great importance in using experimental animals for pain research is their humane treatment. The investigator should

Address correspondence to Krishnaswami Ramabadran, Department 48Q, Abbott Laboratories Building J-23, One Abbott Park Road, Abbott Park, Illinois 60064-3500 or Mylarrao Bansinath, Department of Anesthesiology, New York University Medical Center, 550 First Avenue, New York, NY 10016.

take all precautions to safeguard the rights of the experimental subjects in accordance with the ethical values and prevailing laws. As far as possible it is necessary to minimize the suffering encountered in any experimental design. This is a critical issue, because in order to study pain, it is unfortunately necessary to inflict a certain amount of pain. However, it needs to be remembered that it is not necessary to inflict a degree of pain that can cause intense emotional reactions, indicative of prolonged suffering and agony.

THE INTERPRETATION OF "PAIN" IN HUMANS VERSUS EXPERIMENTAL ANIMALS

Before discussing the methodology of pain assessment it is necessary to know the definition of pain and its relief. According to the subcommittee on Taxonomy of the International Association for the Study of Pain, pain is "an unpleasant sensory and emotional experience associated with actual or potential tissue damage, or described in terms of such damage." This definition clearly distinguishes the pain under examination from, say, the "pain" (grief) of losing a loved one, the "pain" (disappointment) of unfulfilled expectation, or the "pain" (exasperation) of attending an interminable meeting. It also clearly recognizes that pain can and frequently does arise in the absence of noxious (tissue threatening) stimuli, and points to a way of understanding such pain. Pain is not a simple sensation caused by a specific stimulus but rather a complex reaction and experience. Unlike other senses, pain can be induced by a variety of stimuli like chemical, electrical, mechanical, and thermal, etc. Hence the minimal stimulus intensity required to elicit pain cannot be readily specified. The phenomenon of pain has a multidimensional quality because of a wide variety of terms used to describe it, for example burning, throbbing, etc. Pain is also not mediated physiologically by any one tract or nucleus in the brain; it obviously varies widely among individuals and even in the same individual at different times. All these complexities in the qualitative determinants of pain may be due in part to the fact that painful stimuli also influence affective reactions that interact with the sensory components of pain and thus lead to the total pain experience. With these considerations, pain or nociception is viewed as a complex experience, consisting of at least two components, a sensory component referring to the qualitative sensory experience elicited by the stimulus, and a reactive component that refers to the affective and emotional responses which normally accompany painful sensations. This situation remains even more complicated in the case of experimental animals, because it is not possible to obtain verbal reports from them. The term "painful" as used to describe stimuli applied to animals is usually attributed synonymously with the term "nociceptive" or "noxious" which means destructive or tissue damaging in its strictest sense. Alternatively, the term nociceptive or noxious or the term "aversive" is frequently applied to stimuli or events that elicit behavioral responses to avoid stimulus conditions. Another poorly understood aspect is whether animals perceive pain in the same fashion as human subjects, and the pathways transmitting nociceptive messages, brain transmitters, and their receptors in experimental animals are similar to humans. Under these conditions, it would be helpful to define pain in an operational sense based on stimulus conditions and observable responses, and still a better term to use in experimental animals would be "nociception" instead of "pain."

A nociceptive stimulus must be carefully selected since there are a variety of nociceptive stimuli. The techniques employed for the various stimuli differ widely in applicability and limitations. The study of pain should not only be limited to the methods of nociceptive stimulation but also should encompass suitable experimental models for electrophysiological and neurochemical research. None of the current techniques meet all the requirements of an ideal nociceptive stimulus. According to Beecher,[1] the parameters for the nociceptive stimulus must be quantifiable and controlled with precision in order to minimize variability of experimental results due to fluctuations in stimulus parameters; the nociceptive stimulus used should simulate as far as possible naturally occurring stimulus conditions in order to activate the nociceptors involved in pain experience; the stimulus must be easily and frequently repeatable so that the effects can be examined over a number of presentations. This aspect poses problems with nociceptive stimulation for two reasons; first, tissue damage alters the response characteristics of the nociceptors either by sensitization or reduced sensitivity; second, repeated presentations of painful experience leads to anticipatory avoidance learning which interferes with the testing process itself.[2,3,4] The right choice of a stimulus may frequently depend upon the exact nature of the experimental condition and the types of responses that one intends to measure. For example in neuroelectrophysiological applications, it is often needed to provide nociceptive stimuli with a rapid onset in order to elicit evoked electrical activity. This is easily accomplished with electrical or mechanical stimulation but rather difficult to achieve with chemical stimuli. The stimulus should produce the right type of pain relevant to the experimental objective of the investigator. This aspect is very important in analgesic testing when one tries to extrapolate the findings from animal studies to clinical situations, even though it might not be exact. Since different classes of analgesics vary in their mechanisms of pain relief and therefore have different degrees of effectiveness, it is recommended not to rely on only one form of nociceptive test during the determination of analgesic efficacies.

RESPONSES TO NOCICEPTIVE STIMULI IN EXPERIMENTAL ANIMALS

Whereas man can distinguish a wide variety of painful sensation, animals can only display autonomic or somatomotor disturbances indicating that the applied stimulus has taken effect. Somatomotor responses are most commonly employed in the experimental analyses of pain and they vary from one technique to another; they include the tail flick in the mouse or rat, the local contracture of the abdominal musculature in the mouse or rat, vocalization (squeak) in the rat or guinea pig, etc. Some of these responses can be elicited in chordotomized animals and therefore considered to involve polysynaptic reflexes at the level of the spinal cord; however, under normal circumstances they involve long supraspinal neural pathways.[5] Other nociceptive reactions such as licking and jumping in the hot plate test, writhing in response to chemical stimulus and vocalization to an electrical or chemical stimulus, require a high degree of sensory motor coordination. The nature of all such responses, their degree, the concomitant appearance of other autonomic and behavioral symptoms, and therefore the meaning of the whole procedure vary with the parameters of the stimuli and other experimental conditions like freedom or restraint of the animals. Repeated presentation of the nociceptive stimulation modify the responses following local alterations (which might or might not be accompanied by noticeable tissue injury), recruitment, facilitation, inhibition, and/or conditioning. A typical example of such a complex situation is in the case of jumping response in the hot plate technique, which has been well analyzed. Jacob[6] has shown that with single exposure, either a high temperature of the hot plate or a long period of contact of the paws is needed to elicit a jump response; with repeated exposures, jump occurs at much lower temperature of the hot plate and replaces the licking reactions (this phenomenon is called "occlusion"); these phenomena do not occur if the container is low and covered, as the animal learns that it is helpless (this is called "learned helplessness") and therefore they depend on the awareness of the environment; once acquired, these behavioral alterations resist extinction for more than 24 hours (memory). Like the conditioned behavior, these phenomena can at least be partly reversed by submitting the animals to electroshock treatment or by extinction procedure by putting the mice repeatedly on the same apparatus, except not heating the plate. This experiment demonstrates that the thermonociceptive stimulus is still needed to elicit the jump. However, whether the non-heated surface is nociceptive or not is still open to question. It should also be remembered that conditioning might occur even in tests involving simple spinal reflexes like the tail flick, when the animals are repeatedly tested at short intervals or trained.

QUANTITATIVE DETERMINATION OF THE ACTIVITY OF ANALGESICS

Quantification of the activity of drugs differs from one test to another. Depending on whether the nociceptive stimulus is kept constant or not, the methods can be divided into two groups. In the first instance the activity of the drugs is related to the disappearance of responses. One example is the all-or-none procedure. Graded responses can be used either with a conventional scale as in the tail clip method or by measurable features as the response latencies in the hot plate test and the number of abdominal contractions in the writing test. ED_{50} and slope of the regression can also be obtained based on direct quantal treatment. In the second case the activity of the drug is related to the stimulus variations to elicit the responses. These variations might be related to the duration or the intensity of the stimulus or both. When keeping the intensity constant, one measures the reaction time. However, the major disadvantage in this method is to adopt a "cut-off time" which alters the statistical distribution and hence the calculations. If the duration of the stimulus is kept constant, one then measures the thresholds. In reality, the duration is rarely kept constant, since it increases as the threshold is measured. When it is really kept constant, repetition of the stimuli is needed which influences the results and interferes with the objective. When the time-course of the effect is desired, it is necessary to use separate groups of animals instead of repeated exposure of the same animals at a different period of time intervals to avoid the interference of learning by the animals.

SOME COMMONLY USED NOCICEPTIVE TESTS

A great variety of nociceptive tests have been currently used in various laboratories differing from each other by the nature of the stimuli, their parameters, sites of application, the nature of responses, quantitation, and the apparatus used. Objectively, depending upon the nature of the stimulus, they can be classified into chemical, electrical, mechanical, and thermal methods.

Chemically Induced Nociception

A variety of chemical agents have been employed in an attempt to produce a model of pain. One such method is the writhing test. Here the response is unlearned and reflexive in nature. The intraperitoneal administration of noxious chemical substance to mice and rats produces peritoneal irritation which elicits a writhing response. Each episode of writhing is characterized by internal rotation of one foot, sucking in of the belly, elongation of the body, arching of the back, rolling on one side and remaining still, or turning around and circling the cage. Many chemical irritants have been used, which include acetic acid, acetylcholine, alloxan, bradykinin, hydrochloric acid, hypertonic saline, lipoxidase, oxytocin, phenylquinone, and serotonin. Acetic acid is preferred as a chemical writhing agent because of problems of solubility, photosensitivity, and autooxidation with phenylbenzoquinone. Writhing can be abolished by evisceration, intraperitoneal application of procaine, spinal transection at low cervical levels, and ablation of the cerebellum. Midbrain decerebration eliminates arching of the back and decortication does not affect writhing response at all. Writhing is also found to be prevented by electrical stimulation of the periaqueductal gray.[7] One of the major drawbacks is the great variation in individual sensitivity, especially when the number of writhings are considered, which in a period of 20 minutes varies from 0 to 80. Atropine, adiphenine, and dicyclomine are found to be inactive[8] indicating that intestinal spasm is not likely to be involved in the production of writhing.[9] Analgesic activity is considered present, if either the latency to the first writhe is prolonged, or the frequency of writhing is reduced. The writhing test has the advantage of simplicity and sensitivity to all known clinically useful analgesics.[10] Morphine-type analgesics and buprenorphine-like agonist antagonist analgesics are readily detected in this assay. On the contrary, administration of morphine antagonist naloxone has been shown to increase the frequency of writhing in mice and rats[11,12] and this hyperalgesic effect has been shown to be mediated through stereospecific opioid receptors.[12] It should also be remembered that agonist or morphinomimetic actions of naloxone tend to predominate in the writhing test, when high doses of naloxone are employed.[12] The major disadvantage of the writhing test however is its lack of specificity by being sensitive to a number of non-analgesic agents generating false positives[13] and a variety of extraneous variables. Auxiliary tests can be utilized to keep false positives to a minimum. Using a rotarod test in parallel drug-induced motor impairment can be easily dissociated from analgesic effects.

Although either mice or rats could be used in writhing assay, the use of mice is preferred, being more practical and economical. Furthermore, the intraperitoneal injection of chemicals causes less stereotyped and more variable responses in rats than in mice. Animal weight appears to be a significant factor, which, depending on the stimulus, strain, or sex may produce variable results. In general, the writing assay is best conducted in fasted animals. Recently, an automated recording of the writhing response of the test animal using a photoelectric motility monitoring device has been introduced.[14] This device also allows a concomitant measurement of the behavioral locomotor activity of the test animal that negates the observer's subjectivity and other environmental factors. Despite its lack of specificity, a good relationship exists between the potencies exhibited by morphine-type analgesics in writhing assays and their clinical potencies. Since the intraperitoneal injection of bradykinin produces various types of visceral pain in man, writhing produced by this and other noxious substances can be considered as a model nociceptive reaction which accompanies visceral pain. It would be of interest to know which type of nociceptors are involved in this reaction, because primary afferents from the viscera are represented practically by all types of nerve fibers and many of them project to the same spinal cord neurons which are also activated from cutaneous afferents.

Bradykinin-Induced Nociception

Intraarterial injection of only several micrograms of bradykinin and other plasma peptides evokes vocalization and indications of pain in puppies[15] or mature dogs[16] or guinea pigs.[17] Intracarotid (right carotid artery) administration of bradykinin (0.5 to 0.1 μg per kg) in the rat[18] results in dextrorotatory movements of the head, flexion of right forelimb, and an occasional vocal response; repeated administration of bradykinin at regular intervals does not lead to tachyphylaxis (Ramabadran, K., Unpublished data). Once the sensitivity of the individual rat has been established, a dose of analgesic under test is administered by parenteral or oral route and its effects on the bradykinin responses recorded and scored. ED_{50} for the analgesic at the time of peak effect and time course can be plotted. In general, compounds that effectively inhibit writhing in mice are also active against intraarterial bradykinin but the latter may be a little more specific. Although the sensitivity of this technique appears to be the same as in the tail flick, the inhibition of bradykinin response is a sensitive indicator of the analgesic effect of synthetic opioid agonist-antagonists.[19] Interestingly, Satoh et al.[20] observed that in a group of rats in which flexor reflexes displayed tolerance to intraarterial bradykinin (noxious adaptable group), naloxone produced hyperalgesia. This effect of naloxone was absent in those rats which gave a consistent response to bradykinin (noxious non-adoptable group). It has also been shown that the effect of bradykinin is not strictly limited to nociceptors and that other receptors innervated by fast-conducting fibers are concomitantly excited. In spite of all these data, bradykinin and other related peptides are considered as endogenous pain transmitters involved in peripheral mechanisms of pain. Thus they serve as natural stimuli, which when injected intraarterially produce pain which may be similar to clinical visceral pain and thus they are possibly the most appropriate stimulus to use to evaluate potential analgesics. Extreme caution should be exercised, since the pain produced by bradykinin has been shown to be severe in man.

Ethylenediamine Tetraacetic Acid Induced Nociception

Intradermal injection of ethylenediamine tetraacetic acid in the guinea pig evokes vocalization, biting, scratching at the site of injection, and escape behavior[21] and the nociception by ethylenediamine tetraacetic acid was shown to be the result of its cation chelating activity. The suppression of ethylenediamine tetraacetic acid induced responses proves to be a rapid and effective antinociceptive test. Both the opioid agonists and mixed agonist antagonists are active in this test, the non-analgesic drugs are inactive. Similarly, intradermal injection of 50 μl solution of 0.35 N Ammonium hydroxide containing 2.325 percent ethylenediamine tetraacetic acid and 0.837 percent Sodium chloride elicits reproducible short-term behavior in mice typical for a nociceptive stimulus comprising of vocalization, biting, scratching at the site of injection, licking the site of injection, running, and jumping inside the cage.[22] Both the agonists and mixed agonists are effective as analgesics in these two tests. The effect of varying doses of naloxone are not reported.

Formalin Test

Localized inflammation and pain produced by injection of formalin into an animal's paw has been used as a noxious stimulus in several species, for example rat and cat,[23] mouse,[24,25,26] and monkey.[27] In the rat paw formalin test, 50 μl of 5 percent formalin in isotonic saline is injected subcutaneously into dorsal surface of the forepaw or hindpaw. The behavioral responses are either scored[23] or objectively quantitated.[28] In the latter method, the two quantifiable behaviors are (i) flinching/shaking of the paw and/or hindquarters, recorded as the number of flinches per observation period; and (ii) licking/biting of the injected paw, recorded as total licking time(s) per observation period. The flinching is primarily observed as drawing the paw under the body and rapidly vibrating it; since this behavior causes a shudder or rippling motion across the back, it is easy to observe even when the paw itself is not visible. Each episode of shaking or vibrating the paw or each shudder of the back/hindquarters is recorded as a flinch. The effect of formalin on overt behavior is assessed by observing animals in pairs and recording spontaneous flinching and licking for 90 minutes. Results are expressed as mean number of flinches $\pm$ S.E.M. or mean licking time $\pm$ S.E.M. occurring in a given period. Both μ agonists (morphine) and k agonists (PD 117,302) are effective in this test.[28] Naloxone does not produce hyperalgesia in the formalin test, because of an already attained maximal responses (e.g., too high concentration of formalin). Interestingly, the intensity and time-course of pain produced by formalin injection in humans coincides very closely with that defined by behavioral and electrophysiological measures in other species. Observation of formalin-induced behaviors over zero to ten minutes and 20-35 minutes post formalin give reliable and reproducible behavioral scores for early and late phases, respectively. In most instances, quantitation of the early and late phase of the response necessitates use of separate groups of animals. The flinching response in rat is preferred for quantitation of antinociceptive activity, because it is a more spontaneous and robust behavior and is less contaminated by other behavioral influences. However, in mice, licking is the only readily quantifiable behavioral response following formalin injection. The paw licking response in mice is very striking, characteristic, and consistent across time and batches of animals. Central injection of saline alone does not interfere with the nociceptive response in mice. Mice have certain experimental drawbacks vis-á-vis rats, among them quicker

movements that are difficult to observe, a limited behavioral reper-
toire, and a smaller amount of central nervous system tissue available per mouse for neurochemical analysis. However, there are
advantages in using mice, including economy, i.c.v., and i.t. drug
administration without the complexity of surgical cannulation;
smaller requirements of test agents, and a pharmacology often distinct from that of guinea pig, rat, or cat. Ambient temperature plays
a significant role on the licking response in the formalin test in mice
as the licking activity is lower at 20°C than at 25°C.[29] Changes in
paw skin temperature might be responsible for these temperature-
dependent effects. Nociceptive testing should therefore always be
performed in temperature-controlled environments, and the ambient
temperature should be reported.

Electrical Stimulation Methods

Electrical Stimulation of the Tail

Electrical stimulation of the tail through intracutaneous needles
in animals has been shown to produce consistent responses.[30-32]
Three types of pain threshold are determined following the electric
shock applied to the tail;[33] these three nociceptive responses are
considered to correspond to three different levels of integration of
pain within the central nervous system; (a) a low intensity stimulation produces a motor response, called tail withdrawal (which is
low grade spinal reflex); (b) a higher voltage induces a simple
vocalization or squeak, considered as nociceptive reflex involving
the lower brain stem; (c) a stimulation using higher voltage produces vocalization which continues briefly after the stimulation is
terminated and this response is called vocalization after discharge,
which might represent the affective component of the pain response
involving hypothalamus and rhinencephalon. During the testing
procedure, the mouse or rat is kept loosely in restraining cylinders
with its tail free. The electrical stimulation consists of a train of
impulses of 1 second duration with pulse width of 1.5 msecond at a
frequency of 100/second. The voltage is progressively increased by
0.1 volts stepwise until the three thresholds are successively determined. The standard experimental procedure consists of testing the
animals three times at 15 minute intervals during three consecutive

days, after which they become adapted to holding and testing sessions, by the absence of occurrence of any sign of stress such as defecation, teeth chattering, and excessive agitation; they normally remain calm and show no untoward reaction during the insertion of the needles. Thus the threshold values obtained on the third day are considered as baseline values. It has been observed that high doses of morphine were necessary to antagonize vocalization, whereas low doses of morphine could effectively block the vocalization after discharge.[34] Thus it is possible to differentiate between responses which are controlled and integrated at various levels in the pathways mediating nociception and therefore to obtain more information on the neural mechanisms of nociception. Carrol and Lim[34] clearly showed that brain transection between thalamus and midbrain blocked the vocalization after discharge, while higher ablations were without effect. Transection caudal to medulla blocked the vocalization during stimulation as well as vocalization after discharge. These findings reveal that morphine-induced analgesia results from the initial blockade of nociceptive thalamic afferents, followed by blockade of brain stem and spinal neurons. This method is sensitive to morphine-type opioid agonists and anti-inflammatory agents. The opioid antagonist naloxone has been shown to decrease the threshold for vocalization and vocalization after discharge[35] and this naloxone-induced hyperalgesia is only slightly altered in hypophysectomized rats, indicating that the hyperalgesic action of naloxone results essentially from an antagonism of endogenous opioids originating in the central nervous system during the electrical stimulation.

Flinch-Jump Test

The flinch-jump technique provides an alternative to the tail-flick procedure. In this technique, a constant current shock is applied to the grid floor of the rat cage and behavior of the animal is noted.[36-39] The current level is either increased or decreased after each presentation and the order of shock intensity presentation is determined by the ascending and descending series according to the method of limits. As the shock intensity is increased from zero, the first behavior is flinch or crouch or startle; at higher levels of shock intensity, the animal escapes in order to avoid the shock. This test retains the advantages of good sensitivity to opioid and non-opioid analgesics[37] and the measurement of simple responses to nociceptive stimuli. It has been shown that opioid agonists and mixed agonist-antagonists elevate jump threshold without affecting the flinch threshold,[36] and a ceiling effect for raising the jump threshold is observed for nalorphine and pentazocine.[19] This technique allows for free movement of the animal, requires no training and multiple levels of nociceptive stimulation may be studied in a session using a single animal, and stimulus duration and intensity may be more discretely applied. It has been reported that repeated testing at hourly or daily intervals does not appear to alter significantly thresholds of any of the response categories.[37] The response categories derived from observations of the components of motor responses to inescapable foot shock permit the determination of a behavioral response profile. It is likely that each of these response types involves different central nervous system structures mediating rapid responses to foot shock. The flinch response is a locally mediated spinal reflex. A criterion for jump response frequently reported is the elevation of the hindpaws from the grid. Lesions of the dorsomedial tegmentum, septal nucleus, or median forebrain bundle lower the shock threshold for jump response. Vocalization has been observed by only a few investigators in the assessment of responses to foot shock. The vocalization response during foot shock is a vagal reflex, probably mediated at the level of medulla.

Trigeminal Nerve Stimulation

In the rat, direct electrical stimulation of the ophthalmic division of the trigeminal nerve can be used to evaluate morphine-like analgesics.[40] In this method, the threshold current required to evoke a vocal response is measured. More refined electrophysiological approaches to assess neuronal activity in the trigeminal nucleus can also be performed. Biochemical measurements of substance P release from trigeminal slices *in vitro*[41] and electrophysiological studies[42] have shown that nociceptive afferents to this nucleus possess presynaptic opiate receptors.

Shock Titration Technique

Learned or operant responses are a separate category of behaviors from which pain has been inferred in animals. The most common and simplest method involves an animal escaping a noxious stimulus by initiating a learned behavior such as crossing a barrier or pressing a bar. For example, electric shock can be delivered to a grid floor in a cage and the animal can be trained to jump over a barrier partition to escape the stimulus. Usually the latency to escape is measured. A sophisticated version of this type of model is in monkeys, in which there are multiple measures of the animal's escape behavior.[43] Vocalizations are natural pain reactions in animals, and therefore they are used with increasing frequency in paradigms that are employed to assess pain sensitivity. Vocalization in monkeys elicited by electrical stimulation is reduced by systemic morphine (1-2 mg/kg, i.m.).[44]

In the shock titration technique, the experimental subject has operant control over a magnitude of maximum intensity of an applied nociceptive stimulus, usually provided by electrical cutaneous shock with the shock intensity increasing in a step-wise manner unless suppressed by a bar press.[45-49] The test animal therefore determines the tolerated level of shock. The shock intensity is controlled electronically and is programmed to increase by a certain amount at preset intervals. Thus by adjusting its response rate, the experimental animal can control the intensity of the nociceptive stimulus. When stabilization of shock threshold is achieved with daily training sessions for seven to ten days, the animals are then adopted to once-a-week sessions and are tested thereafter at weekly intervals.[49] The animals are continuously observed during the training and testing. There are several measures to be derived from the behavior in this task, and those most frequently reported are response rate, a measure of the maintained stimulus intensity level, or the amount of time spent at each stimulus intensity, the latter has often been employed. As the titration technique can provide a continuous record of escape threshold, it is ideal for studying the time course of analgesic drugs.[50] Once the animals are trained, a number of drugs and a wide range of doses can be rapidly screened for analgesic activity. Some of the disadvantages of this methods are: (a) variations in the size of the increment or decrement produced by the behavioral response have been shown to affect the titration threshold; (b) inconsistencies observed in this technique in the effects of some drugs among different investigators; (c) the aversive thresholds obtained in the titration methods might really reflect avoidance thresholds and that the animals regulate shock intensity to levels which are possible below those actually perceived as nociceptive. The electrical stimulation of the foot seems to have a lesser sensitivity in detecting analgesic activity than the electrical stimulation of gasserian ganglion. This method can detect levels of morphine of the order of magnitude of 0.1 mg per kg.[19] Morphine and cyclazocine[51] but not nalorphine elevate the shock threshold. The elevation of threshold by mixed agonist-antagonist in this procedure requires 100 times more naloxone for reversal than that of morphine.[48] The discrete trial shock titration paradigm is considered suitable for studying the antinociceptive action of opioids and certain non-opioid analgesics but not non-steroidal anti-inflammatory analgesics in the rhesus monkey.[49] It should also be pointed out that some analgesic drugs might produce a number of side effects such as sedation, motor deficits, and mood changes such as euphoria or dysphoria, and these generalized effects themselves influence responding in addition to any perceptual effects that the drugs might have. In rhesus monkeys, morphine reduces sexual behavior, and food reinforced responding in the same dosage range used to produce analgesia. One cannot therefore, with confidence, describe reduction in response to nociceptive stimuli as analgesia, when responses to other type stimuli are reduced.

Tooth Pulp Stimulation

Electrical stimulation of the tooth pulp has been used in antinociceptive tests in several species of animals.[52-65] This test shows promise of being a very useful tool for the study of neurophysiological and behavioral aspects of facial pain. In general, electrodes are implanted in the incisor of the anesthetized animals (rats, dogs) or in animals supplemented with analgesics (rabbits). After a recovery period that varies from 24 hours for rabbits to seven to ten days for rats and dogs, animals are placed in experimental chambers and allowed to acclimatize to the test environment. Electrical stimulation usually produces a licking/biting/chewing or head flick response. The threshold voltage or current that causes the response is determined and is considered as the criterion for pain threshold. Considerable controversy also exists whether receptors in the tooth pulp have an exclusive role in mediating nociception. The findings that the innervation of the tooth pulp consists exclusively of fibers in the A-δ and C-fiber range, indicate that they conduct nociceptive information along these fibers. Non-nociceptive sensations might be explained by the spread of electrical stimulus outside the pulp to gingival and/or periodontal tissues, which are also innervated by fast-conducting fibers, supplying to mechanoreceptors.[66] The risk of current spread is highest with monopolar stimulation, using an "indifferent" electrode in contact with the skin in the hand or ear lobe, etc.; under these circumstances, the receptors other than those in the tooth might be excited by the electrical stimulation.[67] Furthermore, if the tooth surface is not completely dry, moisture can cause the current to be shunted along the surface of the tooth to gingival tissues and result in reduced current flow through the pulp with the resultant stimulation of non-pulp receptors. These problems might partially be solved by using bipolar electrodes. An appropriate stimulator can then be used to stimulate the pulp. Tetanic stimulation of the tooth pulp produces a nociceptive reaction in an awake cat which can be graded according to the intensity of the stimulation.[68] Oliveras et al.[68] suggest a four-point scale to describe the nociceptive behavior. At low voltages, only jaw opening can be elicited; further increases in voltage evokes hyperextension of the head; further increase in voltage results in jaw opening and rotation of the head; finally maximal stimulation produces scratching of the tooth. According to Mason et al.,[69] the suppression of the jaw- opening reflex evoked by tooth pulp stimulation is considered to be an accurate index of analgesia, and therefore the tooth pulp test provides a sensitive, reproducible, and clinically predictive method for evaluating the antinociceptive actions of opioid agonists, mixed agonist-antagonists, and centrally acting non-opioid analgesics. Peripherally acting analgesics are inactive in this test. Antinociceptive profiles of μ and k opioid agonists can also be studied in the tooth pulp stimulation procedure in awake, freely moving rats; the nonsteroidal anti-inflammatory drugs, aspirin and zomepirac, are effective analgesics but the dose response curve slopes for these compounds are lower than that of morphine-type analgesics and the morphine antagonist, naloxone produces no significant changes in threshold responses. However, naloxone reverses the threshold increases produced by morphine and ethylketocyclazozine but not aspirin.

Mechanical Stimulation

Selective stimulation of the mechanoceptors is possible by application of high pressure. However, when applying high pressure, it is not possible to avoid stimulating low threshold mechanoceptors. The disadvantages of mechanical methods are that mechanical stimuli might produce receptor damage and therefore repeated application of the stimuli does not always elicit reproducible results. Even though mechanical stimulation can be easily applied, fine control and measurement of stimulus parameter are difficult to achieve without expensive equipment. Various adaptations of the Von Frey technique, using hair or nylon strings of different diameters and lengths, are the most common methods of quantitative expression of pressure. The hair is attached to a lever and the intensity of pressure is measured as weight per mm.^{2} The threshold for evoking nociception is usually above 40 g per mm.^{2}

Tail Clip Test

The tail clip test introduced by Haffner[70] is probably the most universally used and considered as a crude technique in the neuropharmacological blind screening, in the place of, or in addition to, the pinching of the forepaws. The animals generally used are the mice or rats, although occasionally cats and monkeys are used. Bianchi and Franceschini[71] studied this method extensively using a "constant" stimulus by means of an artery clip, the arms being enclosed in a thin rubber tube. The artery clip is applied to the base of the mouse's tail for 30 seconds. The animal reacts by locating the clip continuously with attempts to dislodge the clip by biting it. Therefore it can be inferred that the response of oriented biting involves higher centers in the central nervous system. It is equally important to adjust the pressure exerted by the clip so that it is just sufficient to cause all control mice to respond by attempting to dislodge the clip. The animals are first screened and those not attempting to remove the clip after 15 seconds are excluded. Morphine and test drugs are administered and the test repeated at 30, 60, 90, and 120 minutes after drug administration. Animals not responding after 30 seconds are scored as positive. A cut-off time of 30 seconds is used. An ED_{50} readily is calculated from the data. The ED_{50} of morphine varies depending upon the duration of the cut-off time. If the cut-off time of 30 seconds is used, the ED_{50} is found to be 5.7 mg/kg s.c. in mice, whereas with cut-off time of 10 seconds, the ED_{50} is reported to be 7.1 mg/kg s.c. in the same species. The main problem in performing this test is in standardizing the point of application of the clip to the tail of the animal. A convenient method for minimizing the variation arising from this factor is to clearly mark the point of application of the clip on the tail and on the clip. Repeated application of the clip to the tail can cause swelling which might alter the reaction profile during the course of the test. Perhaps the greatest source of variation is in the opening tension of the clip. Even though this test appears to be a crude technique, it has a wide application as a basic screening test in the evaluation of morphine-type analgesics. Opioid analgesics cause the mice to be indifferent to the clip. Clips can also be applied to other parts of the body such as ears, paws, and toes, which may be more convenient for bigger species than mice.

Caudal Compression Test

In the caudal compression test,[72] the rats are usually restrained. The rats are positioned so that the pressure is applied about 1 cm from the base of the tail or 2.5 cm from the tip of the tail. Threshold pressures are measured with a simple two syringes connected by means of a non-elastic, flexible plastic tubing filled with a fluid and the side arm in the tubing is connected to a manometer. The manometer reading is taken when the rat responds to pressure, first by struggling, then by vocalization. Vocalization is normally regarded as the most specific central indicator of nociception in animals. The tip of the tail is best for obtaining a prompt response. In subsequent determinations following morphine or test drug administration, the pressure is applied to the same place on the tail. A mechanical drive is convenient for moving the second syringe so that the pressure increases uniformly at 20 mm of mercury per second and can be instantly released following the squeak. The problem of standardizing the point of application of the pressure on the rat tail is one of the main difficulties of this test.

The tail compression test can also be used in mice, using a mechanical pressure of 10 mm Hg per second[73,74] The level of pressure in mm Hg that evoked biting or an aversive response is noted. Only mice responding behaviorally to a tail pressure of 40-50 mm Hg are selected. A value of 100 mm Hg is normally used as the cut-off pressure to avoid tail tissue damage. The antinociceptive activity for each mouse can be calculated according to the formula:

$$\% \text{ of antinociception} = (P_2 - P_1/100 - P_1) \times 100$$

where P_1 is the responsive pressure before drug injection (mm Hg) and P_2 is the responsive pressure after drug injection. At 15, 30, and 60 minutes following morphine administration, tail pressure thresholds are determined. The pressure thresholds can also be determined after i.c.v. administration.

Paw Pressure Test

Hyperalgesia, a feature of inflammatory pain, decreases the pain threshold and is caused by the sensitizing effect of mediators such as prostaglandins on nociceptors.[75,76] Administration of brewer's yeast into the sub-plantar area of the rat paw results in acute inflammation and hyperalgesia.[77] Two hours after the yeast administration, the pain threshold in the inflamed and noninflamed paw can be determined by applying increasing pressure on the paw until a withdrawal or escape response or vocalization is obtained.[78] Depending on the device, the pain threshold is expressed as g/force or as mm Hg. The original Randall-Selitto method can be modified so that a constant pressure, usually 20 mm Hg, is applied and the reaction time or latency to pain response (escape, struggle, or vocalization) is measured.[79] The animals are then randomized and divided into groups so that the pain threshold or reaction time is similar in each group. An analgesic drug or vehicle is subsequently administered, and the pressure is reapplied at appropriate timings. Opioid analgesics generally increase the threshold of the inflamed paw as well as the noninflamed normal paw, whereas the nonsteroidal antiinflammatory drugs act only on the inflamed paw. The potencies of nonsteroidal antiinflammatory drugs vary widely from laboratory to laboratory, probably due to the criteria set by the individual investigators and also due to subjective observations as well as animal handling by operators. Despite the variability, the ED_{50} values of nonsteroidal antiinflammatory drugs in this test are reported to be predictive of human doses.[80] Other noxious agents commonly employed to induce inflammation in rats are carrageenan, Freund's adjuvant and PGE_2. Usually measurements of the pain threshold of the inflamed paw can be taken two to three hours after the intradermal injection of carrageenan, or two hours after PGE_2, or 18-24 hours after Freund's complete adjuvant. In some laboratories, Freund's complete adjuvant is injected into the proximal part of the rat tail and hyperalgesia is demonstrated by the rat's squeaking response, when the pressure is applied at the site of injection.[81]

The paw pressure test can also be used in guinea pigs, using a load of 75g with an analgesymeter.[82] As with the rat paw pressure test, the end point is determined subjectively, and it is essential for each complete test to be performed by a single operator who is 'blind' to drug treatment. The typical and consistent end point response is the withdrawal of the paw. Distinction among μ and k opioids can be made using both rat and guinea pig paw pressure test. K opioids are more potent analgesics in the guinea pig than in the rat paw pressure test, whereas morphine-type analgesics are equipotent in the two tests. These differences in rat and guinea pig might possibly be related to the increased densities of k binding sites in the guinea pig brain, and the relative lack of k binding sites in the rat brain.[83] Interestingly, administration of the opioid antagonist, naloxone, alone into inflamed paw has been shown to antagonize the inflammatory pain produced by the intraplantar injection of carrageenan in rats.[84] This paradoxical analgesic action is mediated by stereospecific opiate receptors and is explained due to a peripheral action on cutaneous opiate receptors and/or local production of morphinomimetic metabolites locally formed by the dealkylation of naloxone.[85]

Thermal Methods

Heat as a nociceptive stimulus has the advantage that it does not stimulate the mechanoceptors. The heat stimulus is natural, relatively easy to control, and under certain conditions elicits nociception without tissue damage. Two reactions, the skin twitch and the escape or withdrawal, occur at average skin temperatures of 45 to 46°C, and 51 to 52°C in experimental animals. Conducted heat stimuli could be easily applied to a moving subject as in the hot plate technique or to immobilized animals in the radiant heat methods. Changes in the skin temperature by drugs should be carefully monitored, because they could result in spurious alteration of pain thresholds, and/or misleading results with regard to the mechanisms underlying the production to either analgesia or hyperalgesia.[4] The tail is an important organ in the thermoregulation in mice and rats, and in response to elevated ambient temperature, both tail blood flow and tail skin temperature are increased.[86,87] It has been reported that the mean cutaneous tissue temperature in the tail at which a tail reflex is evoked is relatively constant (approximately 45°C). It has been suggested that the change in thermal threshold needed to evoke a tail flick reflex is an important measure of altered tail flick reflex control. At higher initial skin temperature, this threshold is presumably reached more rapidly leading to reduced tail flick latencies. The difference between the estimated threshold and normal skin temperature in rodents is in the order of 12 to 18°C and a change in skin temperature would therefore be expected to produce significant changes in tail flick latency. In fact, a significant negative correlation between tail skin temperature and tail flick latency in mice has been demonstrated.[89] Apart from pharmacological and surgical treatments, other manipulations which alter tail blood flow or thermoregulatory mechanisms may be expected to produce apparent hyperalgesia or antinociception in the tail flick test. Interestingly, the tail skin temperature can also be influenced by the handling of the animals. There is experimental evidence to show that non-noxious stress affects the body temperature and responsitivity to nociceptive stimulation in rats.[90]

Tail Flick Test

The tail flick procedure of D'Amour and Smith[91] has become standard screening procedure for testing analgesics in both mice and rats. Although the basic elements of the apparatus have not changed since the work of D'Amour and Smith, the sophistication of the apparatus has been improved. The rodent is confined with the help of a cloth towel or a plastic holder so that it is relatively immobile. The use of cloth over the animal tends to be less stressful and helps to keep the animal calm. One must be careful not to suffocate the animal or hold it so tightly that it cannot flick its tail. A radiant heat from an electric source (1000 W) is focused on the marked end of the tail. The intensity of the light is varied so that the tail flick latency can be carefully selected depending on the experimental protocol. The tail flick latency can be accurately and automatically recorded with the help of a photocell. A suitable cut-off time can be used. Once the tail flick has occurred, the radiant heat stimulus is simultaneously terminated to avoid tissue injury. This technique has been modified by employing a red-hot wire 3 mm from the tail. The same basic system can be used in dogs and guinea pigs where the end point is a skin twitch. Generally two pre-drug control reaction times are determined 15-30 minutes apart. Each animal serves as its own control. The drug is then administered and the tail flick latency measured at 15-30 minute intervals. An ED_{50} can be calculated from the number of animals that do not respond within the cut-off time (ten to 20 seconds, depending on the needs of the testing protocol). The analgesic activity can be calculated as percentage of maximum possible effect (%MPE), using the formula

$$\%MPE \ = \ \frac{[\text{Postdrug latency}] - [\text{Pre-drug latency}]}{[\text{Cut-off latency}] - [\text{Pre-drug latency}]} \times 100$$

The detailed procedures are described.[92] At low intensities, most opioid agonist antagonists with marked antagonistic activity like cyclazocine are not detected.[93] The opioid antagonist, naloxone, does not affect the reaction time in mice, whether the control tail flick latencies are short or long.[94,95] The strength of the stimulus does not appear to play a significant role and may not be the sole reason for the lack of effectiveness of opioid agonist antagonists in this test. The rats and mice used in this technique should be young so as to minimize the heat insulating effect of keratinization in the tail. The color of the skin in the tail area should be consistent to standardize the heat absorption.[96] The locus of tail stimulation in the tail flick assay is an important parameter in determining the analgesic effects, as the distal tail section has been reported to be more sensitive to the analgesic effects of morphine than the proximal sections.[97] Tail vein injections should be avoided in the tail flick procedure. There is also evidence to show the existence of diurnal variation of tail flick to nociceptive stimulation.[97a] It is also necessary to control the skin temperature of the tail, when using radiant heat in the tests of analgesia.[98-100] The animals are restrained during this procedure, and hence are stressed; this restraint stress might itself activate endogenous opioid and/or non-opioid pain inhibitory systems which interfere with the test.[101-103] The tail flick response does not involve a high degree of sensory-motor coordination, as the response varies little between intact and spinal animals. Although it is a low-grade spinal reflex, the tail flick response is normally under the physiologic control of higher centers. In rats transacted at high thoracic levels, tail flick reflex still persists, and is even more exaggerated than usual, probably as a result of abolition of descending inhibition from higher centers.[100] The tail flick latency may become shorter than normal, partly due to peripheral vasodilatation following spinal transection. In the case of mice it has been shown that if one transects the spinal cord, and then exposes them to nociceptive stimulus of the tail flick procedure, the mice still flick their tails.[104] Morphine is able to produce a 30 percent decrease in response latency in the spinal mice. These data suggest that although the spinal reflex is a factor in the tail flick test, there is some supraspinal influence that is blocked by morphine in nonspinal animals and cannot be blocked by morphine in spinal mice. Mayer et al.[105] have reported that electrical stimulation of the paraventricular gray areas of the brain cause an increase in the latency of the tail flick test. In addition, it has also been shown that the intraventricular injection of morphine causes an increase in tail flick latency. These results support the hypothesis that supraspinal mechanisms are also involved in the actions of opioids in the tail flick procedure. Training or repeated testing at short intervals might also affect the tail flick latency under treated conditions. Under these circumstances it is very difficult to distinguish whether the observed modification is due to the measure of repeated testing alone or due to the effect of the drug treatment. Two other factors which might also influence the tail flick latency are the cut-off time and the effect of the drug treatment on body temperature.

The rodent tail flick assay is commonly used to assess opioid analgesics, although other centrally acting drugs have shown activity in this test. Aspirin-like peripherally acting analgesics are usually inactive or show very little activity at toxic doses. Even though the tail flick is commonly employed, this test is not as sensitive as some of the other tests which are used for opioid analgesic screening. The ED_{50} for morphine in the tail flick test using the s.c. route is approximately 6-8 mg/kg in mice or rats.

Tail Immersion Test

A variation of the tail flick test is the tail withdrawal assay, also called the tail immersion assay, in which the rodent's tail is immersed in hot water at 45-65°C.[106] From a practical standpoint, rodent tail immersion test has many advantages of a good assay procedure. The mice or rats are held by gently wrapping them in the chux and the terminal 3 cm of the animal's tail is immersed. The nociceptive endpoint is characterized by a violent jerk of the tail.[107] Baseline latencies are determined twice, 5 minutes apart and averaged to give a single predrug latency. In the morphine-treated groups, tail withdrawal latencies are measured before (basal) and 30, 60, 90, and 120 minutes following drug injection. In the naloxone-treated animals, latency is measured before and 15 and 30 minutes after injection. In order to minimize the tissue injury due to repeated exposure of the animal, a cut-off time of 15 seconds is imposed. Analgesia is expressed as percent of the maximal possible effect (%MPE).

$$\%MPE = \frac{[\text{Postdrug latency}] - [\text{Pre-drug latency}]}{[\text{Cut-off latency}] - [\text{Pre-drug latency}]} \times 100$$

With water temperature of 50 ± 0.2°C, and ED_{50} of 5 mg/kg (s.c.) of morphine is reported.[107] This method shows about the same sensitivity as the tail flick to radiant heat test. Young animals should be used, as keratinization of the tail increases with age and this alters the reaction profile. Injections in the tail should be avoided. Special attention needs to be given to the method employed to restrain the animal while testing the tail flick latencies in this test.[107] Although this assay is specific for opioid agonists, by varying the water temperatures, which mimics differences in pain intensity, this assay allows the detection of analgesic activity of the opioid agonist antagonists as well as milder analgesics of the nonsteroidal antiinflammatory type.[108,109] The opioid antagonist, naloxone (1 and 3 mg/kg , s.c.), shortens the tail withdrawal latency in the tail immersion test,[107] suggesting that the tail flick response in this test is tonically inhibited by the endogenous opioid systems. Interestingly, at a stimulus intensity of 45°C, naloxone increases the tail flick latency, it decreases latency at 50°C, and does not alter the latency at 55°C. The analgesic effect of naloxone observed at 45°C appears to be a manifestation of its morphinomimetic properties,[3,83,107,110,111] while lack of any effect at 55°C may be due to very short tail flick latency at this stimulus temperature, which impose practical restriction of measurements.

Another variant of the hot water tail immersion test is the cold water tail immersion assay.[112,113] In this model, the noxious stimulus is cold water (–10°C), mixed with a 1:1 solution of ethylene glycol. The water temperature is maintained using a circulating bath. The rats are held by gently wrapping them in chux and the lower half of their tails immersed in the cold water. The nociceptive threshold is taken as the latency until the rat flicked its tail. Animals failing to respond after 60 seconds of tail immersion should be discarded. Latencies are measured 30 minutes before and 30, 45, and 60 minutes after the injection of morphine and other drugs. Percent of maximal possible effect (%MPE) for each animal at each time is calculated using the formula similar to that described for the hot water tail immersion test. Morphine produces a significant effect at an ED_{50} of 5 mg/kg (s.c.) 30 minutes post injection. The cold water tail immersion test can assess the analgesic activity of morphine-type agonists and buprenorphine-type mixed agonist antagonists and also cyclooxygenase inhibitors. It has been claimed to be highly sensitive to kappa opioid analgesics,[114] which needs to be confirmed.

Hot Plate Technique

The hot plate technique originally devised by Woolfe and MacDonald[115] is another frequently used nociceptive test using the rodents. In this test, a thermonociceptive stimulus is applied to the paws of the animal and the animal can move freely inside the container. Several modifications of this test have been described, differing from each other by the temperature of the hot plate, either constant or variable, the container or enclosure (behavior of the rodent depending upon its size, height, and diameter), the number and duration of exposures, the cut-off time, analysis of results, etc. The most commonly used apparatus is that of Eddy and Leimbach[116] which gives reproducible results. The temperature of the hot plate is kept constant by boiling fluid or a combination of two fluids. The container is a glass cylinder open at both ends. The responses of the animal can be easily observed through the glass. The animal is gently introduced onto the hot plate. The reaction times are measured. The responses are agitation, rapid withdrawal of the paws, "dancing" (not used in analgesic measurements), licking of the forepaws and/or hindpaws, and jumping (simple jump, adjusted jump, or jumping off the hot plate). The most commonly used responses are licking and jumping. These responses occur in almost all animals, can be scored reliably, and are considered equally valid with respect to predicting analgesic efficacy. The experimental animal might be a mouse or rat. A complete analysis of the action of any substance can be made after measuring the various reactions on the hot plate and evaluating the behavior of the animals. For example, the effects of morphine on the licking and jumping responses are not identical; it has been observed in the past that the licking response shown by animals on the hot plate was less sensitive to morphine than the jumping reaction. These differences indicate that opioidergic pathways are predominantly involved in a complex response such as jumping, but they are also involved to a lesser degree in simpler ones such as licking. Moreover, some drugs affect the licking and jumping in different ways. For example, atropine-like drugs inhibit selectively the effects of morphine on the jump through a central mechanism,[117,118] and d-Lysergic acid Diethylamide (d-LSD) preferentially affect the licking, and other hallucinogens affect both responses at the same time.[119] It has been generally accepted that low doses of naloxone produce hyperalgesia as measured by decrease in licking and jumping latencies in mice and rats in the hot plate test[3,83,110,120,121] and this has been confirmed.[122-124] In order to observe the hyperalgesic effect of naloxone, it is necessary to use test-naive animals and a limited number of pre-exposures; when the number of pre-exposures has been sufficient to decrease the jump latencies to very low values, naloxone tends to increase the jump latencies.[3,4,83,125] Many investigators determine the baseline nociceptive threshold by repeated testing of the animals in order to stabilize them but concomitant learning occurs which interferes with this method.[126-131] The licking reactions are also modified by naloxone provided the temperature of the hot plate does not exceed 50°C. At a hot plate temperature of 55°C, the licking latency has already attained the lowest values, and under these conditions they could not be further decreased by naloxone. It has been already shown that the jump response is a very complex one, involving awareness to the environment even on the first exposure, because facilitation occurs on the second exposure. This type of learning could be inhibited, if the environmental conditions relating to the first exposure are inadequate. Despite the facet that it is difficult to detect the analgesic effect of many agonist antagonists by the conventional hot plate temperature at 55°C, O'Callaghan and Holtzman[132] reported that by reducing the temperature of the hot plate to 49.5°C, an analgesic action of pentazocine, nalorphine, cyclazocine, and levallorphan could be detected in the rat and this analgesia could be antagonized by naloxone. Although this test is probably less sensitive than the tail flick, it relies on a more complex and integrated response of the experimental animal, thus presumably reflecting processes occurring at a higher level of the central nervous system. Detection of non-analgesic drug activities in the hot plate test can be minimized by use of a hindpaw lick or jump endpoint. It is most commonly used to assess analgesic effects of opioids and non-opioids hyperalgesia produced by opioid antagonists, amine antagonists, and other drugs.[3,83] The analysis of genetic variations might also be helpful in the elucidation of various aspects of the regulation of nociception.[133] For example, CXBK strain of mice are poor in cerebral opiate receptors and/or their ligands and they respond fast in the hot plate test at 55°C.[134] On the other hand, CXBH strain of mice appear to be rich in cerebral opiate receptors and/or their ligands and they can withstand nociception for a relatively longer period in the hot plate test and therefore they are slow responders to the thermonociceptive stimuli.[135] Interestingly, CXBK and CXBH strain of mice appeared to be deficient in kappa receptors and/or their ligand levels relevant to thermonociception, as administration of U-50488H, (–)bremazocine, and Mr 2266 in these mice does not produce either prolongation or a reduction in jumping latencies. At present it appears that mu, kappa, and delta types of opiate receptors are involved in the regulation of thermonociception.[83,136-139]

Other Methods

Ultrasound Procedure

Ultrasound generates a nociceptive reaction of a different character than that arising from electrical, mechanical, and thermal stimulation.[140,141] Whether this stimulus more closely resembles pathological pain is not clear at present. However, when applied to the tails of rodents, it causes reliable response that is sensitive to morphine-type analgesics.

Laser-Induced Nociception

A new advance for obtaining radiant heat nociception is through the utilization of CO_2 laser-induced heat. The laser beam is focused on a nonhairy surface of the rodent skin, such as the ear, paw, or tail, and behavioral responses are recorded.[142] The method offers many advantages: This procedure can be carried out in unrestrained animals in their home cages, thus there are no environmental influences on the test results. Second, the intensity of the stimulus is kept constant, and the threshold remains stable. Third, this method does not produce tissue damage. Morphine-type agonists are active in this test, whereas the action of antiinflammatory analgesics remains to be reported.

MODELS FOR CHRONIC PAIN

Traditional concepts, technical problems, and ethical considerations remain as barriers to the development and utilization of animal models of chronic pain. Because of those constraints, theory and therapy of chronic pain have progressed relatively little. Chronic animal models would be highly valuable for testing the hypotheses concerning causative mechanisms, development of effective therapies, provisions for new strategies, and generation of novel concepts. The nociceptive methods so far described for examining the effects of analgesics are inadequate for the study of chronic and paroxysmal pain.

Dorsal Root Rhizotomy

Self-mutilation which occurs after rhizotomy or peripheral neurectomy in laboratory animals represents a model of deafferentation or "phantom" pain in humans. Rats self-mutilate the limb ipsilateral and sometimes also contralateral to a dorsal root section.[143-144] The self-mutilation occurs as a result of painful or abnormal sensations attributed to the denervated limb. These abnormal sensations are caused by changes in the activity of deafferentated cells in the dorsal horn.[143] Autotomy is seen as an effort to suppress a region of referred pain and is comparable to the pain caused by denervation hypersensitivity associated with brachial plexus avulsion in humans.[145] Dorsal rhizotomy is typically followed by a scratching and chewing of the ipsilateral forelimb. Scratching results in a progressive denuding of the skin, while chewing leads to various degrees of amputation of the digits and foot. While chewing is usually restricted to the denervated area, scratching may occur on innervated skin adjacent to the deafferented region. Scratching may also occur in the contralateral "mirror zone," in an area symmetrical to the opposite partially afferented region. Autotomy typically occurs within the first month after rhizotomy with a mean onset between ten and 15 days. The relationship between autotomy and abnormal neuronal activity is supported by observations that self-mutilation following deafferentation is blocked by antiepileptic drugs such as carmazepine and sodium valproate.[150]

Autotomy Following Sciatic Nerve Section

Another model that is considered to produce persistent pain involves the complete sectioning of the sciatic nerve in the rat with encapsulation of the cut and resulting neuroma formation. After approximately two weeks, the animals engage in self-mutilation of the denervated area, beginning with chewing the toenails, and followed by a progressive degree of amputation of the digits and foot.[151-154] Mutilation of the denervated extremity following sciatic nerve section has been suggested to reflect abnormal or unpleasant sensations as a result of the nerve injury. Increasing evidence has accumulated to support this behavior as an indicator of neuropathic pain, although controversy exists.[149,155] Chronic infusion of morphine (240 μg/day) into the intrathecal space around the lumbar spinal cord, starting immediately after sciatic nerve section, is reported to reduce the incidence of autotomy in rats, while the severity is increased and the onset reduced following intrathecal infusion of the long-acting opioid antagonist, naltrexone (μg/day).[156] In contrast, autotomy is reduced by continuous s.c. administration of the universal antagonist, naloxone (80 μg/hour for five weeks, or 800 μg/h for two weeks), via an osmotic pump.[157] Considering the possibility that high doses of naloxone produce at least some morphinomimetic actions,[3,83,85,96,110,111] the reduction of autotomy is not surprising. If the view that autotomy occurs as a result of chronic discomfort or pain following nerve injury is correct, then morphine injection should be effective in alleviating neuropathic pain. In fact, this concept has been confirmed in rats after dorsal root rhizotomy.[158] However, there is little data in the clinical literature concerning the analgesic effect of morphine against chronic neuropathic pain in patients.

A more suitable and reliable model is the one developed in rats, in which a partial deafferentation is produced by gentle ligation of the sciatic nerve with resultant demyelination of the large fibers and destruction of some unmyelinated axons.[159] This model is considered to mimic the clinical condition of painful neuropathy with evidence of allodyna, hyperalgesia, and spontaneous pain.

Adjuvant-Induced Arthritis

Other methods have been developed to mimic human conditions of chronic pain. The adjuvant-induced arthritis in rats is a widely used model for the evaluation of drugs in chronic pain.[160,161] Rats are injected intradermally with a suspension of dead mycobacterium butyricum or tubercle bacilli into the tail. Hypersensitivity to pain develops within a few hours and is stable for testing at 18 to 24 hours. By the tenth to eighteenth day, arthritis is established in the hindpaws. Pain is inferred from scratching behaviors, reduced motor activity, weight loss, vocalization following pinching of the affected limbs, and a reduction in these responses following the administration of morphine and related drugs.[162-164] The adjuvant-induced arthritis model in the rat is primarily useful for the evaluation of nonsteroidal antiinflammatory analgesics, although it is equally sensitive to morphine-type analgesics. Morphine-type analgesics act immediately, whereas there is usually some delay in the case of nonsteroidal antiinflammatory analgesics. Although there is controversy regarding the validity of the adjuvant-induced arthritis model, it is suggested that arthritis in rats is associated with chronic pain, stress, and deprived states.[161] The assessment of pain in this model still poses technical problems. Until true quantitative measurements can be made, the results are subjected to subjective bias based on observations of behavioral responses. Furthermore, it should be noted that this is a systemic disease of the animal that includes skin lesions, destruction of bone and cartilage, impairment of liver function, and a lymphadenopathy.[165] These systemic lesions make it more difficult to relate the animal's behavior with pain, as opposed to generalized malaise and debilitation. The occurrence of central nervous system changes associated with modifications of immune function also question the validity of this model. Additionally, arthritic pain is not the only type of chronic pain observed in clinical conditions.

A more recent model of arthritis for pain research is the one in which arthritis is produced by the injection of sodium urate crystals into the ankle joint of the rat.[165,166] The arthritis is fully developed within 24 hours. These rats reduce the weight placed on the treated hind limb and elicit careful movements of the affected paws. There are no signs of systemic disease in the urate arthritis model other than joint pathology secondary to tissue edema and the infiltration of polymorphonuclear leukocytes.[165]

All these chronic pain models, with the aim of mimicking the human conditions of chronic pain, lead to the production of pain, which the animals cannot control. Therefore it is recommended that investigators assess the level of pain in these animals and alleviate this pain by suitable analgesics, but at the same time without interfering with the objective of the experiments.[167] Pain in these studies can be inferred from ongoing behavioral responses such as drinking, feeding, sleep/wakefulness cycle, grooming, and other social behaviors. Significant deviation from normal ongoing behavior is indicative of severe and possibly intolerable pain.

CONCLUSION

In all these tests it should be emphasized that motor impairment or incoordination must not be confused with real antinociceptive activity. For example, prolongation of jumping latencies in the hot plate test after the administration of high doses of 5-Methoxy-N,N-dimethyltryptamine and cyproheptadine appear concomitantly with motor disturbances such as ataxia, reduced muscular tone, and flat posture, and no real antinociceptive actions are indicated by vigorous paw shakes of these animal on the hot plate.[168] Commonly, the selectivity of the analgesic effect is determined separately by observing the motor performance on other tests such as the rotarod test or inclined screen test. As an alternate, analgesia could be measured in such a way that determination of the presence or absence of analgesic activity is independent of motor performance. One such technique involves the use of an analgesic drug as a discriminative stimulus in rats with chronic arthritis. Despite the difficult tasks of assessing pain, anxiety, and euphoria in experimental animals, the characterization and evaluation of analgesics remains relatively easy at the present time. Detection of analgesic action is efficient in most antinociceptive tests. Nociceptive tests for assessing the effect of opioid antagonists and antipyretic and anti-inflammatory agents should be carefully carried out. Nevertheless, newer techniques are evolving, which under appropriate conditions could predict the analgesic or hyperalgesic actions of these compounds. The number and nature of tests to be employed rely on the expertise of the pharmacologist. Experiments with mice and rats have definite advantages because the activities of reference drugs are well established and documented in these species, with tests involving both simple and complex responses, with a high degree of central processing. Activity of a compound must be reconfirmed by a battery of several nociceptive tests which vary widely in the site and mode of stimulation. The nature of the test should also depend on the economy in time and the number of animals used. Since animals are monitored for their behavioral responses (licking, biting, squeaking, jumping), reflexes (tail flick, tail jerk), or muscular contractions, such subjective assessment varies from laboratory to laboratory. Therefore, besides the question of simplicity and expediency, the investigator must ensure for the reliability and reproducibility of the results for rigorous statistical analysis. In addition to the antinociceptive tests, tests should also be included to evaluate the dependence liability of morphine-like compounds.[169-171] One should also be aware of the limitations of drawing analogies between animals and man in the field of pain, and one should not extrapolate too much from animals to human subjects. Another important strategy for studying pain in experimental animals is to carry out observations in their natural habitat. *In vitro* methods might also be used to help the experimenter in the rapid screening of analgesics to understand fully the mechanisms involved. In the development of potential analgesics, attention should also be focused on side effects and/or toxicities. Increased knowledge in the development of biochemical, behavioral, anatomical, and electrophysiological analyses of different types of nociceptive stimuli and of the responses might lead to more reliable ways of relieving pain in the future. The availability of simple, rapid, and reliable assays has already led to the discovery of numerous analgesics effective in mild to moderate pain. Despite all these efforts, a safe analgesic devoid of addiction liability remains to be identified. Future progress of development of new analgesics depends on animal research. All investigators in pain research bear a high and continued responsibility in the proper treatment of animals that participate in pain research.

REFERENCES

1. Beecher, H. K., The measurement of pain, Pharmacol. Rev., 9, 59, 1957.

2. Ramabadran, K., Importance of opioidergic and aminergic systems in the regulation of nociception: Demonstration and analysis of hyperalgesic effects of their modifiers, PhD Thesis, University of Paris VI, 1980.

3. Jacob, J. J. C. and Ramabadran, K., Role of opiate receptors and endogenous ligands in nociception, Pharmacol. Ther., 14, 177, 1981.

4. Ramabadran, K. and Bansinath, M., A critical analysis of the experimental evaluation of the nociceptive reaction in animals, Pharm. Res., 3, 263, 1986.

5. Irwin, S., Houde, R. W., Bennet, D. R., Hendershot, L. C. and Seevers, M. H., J. Pharmacol. Exp. Ther., 101, 132, 1951.

6. Jacob, J., Some effects of morphine on adaptive and learning behavior, In *Psychopharmacological Methods,* Votava, Z., Horvath, M. and Vinar, O., eds., State Medical Publishing House, Prague, 1963, 70.

7. Giesler, G. J. and Liebeskind, J., Inhibition of visceral pain by electrical stimulation of the periaqueductal grey matter, Pain, 2, 43, 1976.

8. Sigmund, E. R., Cadmus, R. and Lu, G., Screening of analgesics, including aspirin type compound based upon the antagonism of chemically induced "writhing" in mice, Proc. Soc. Exp. Biol. Med., 95, 729, 1957.

9. Jacob, J., Evaluation of narcotic analgesics, In *Methods in Drug Evaluation,* Mantegazza, P. and Piccinini, F., eds., North Holland Publishing Company, Amsterdam, 1966, 278.

10. Wood, P. L., Animal models in analgesic testing, In *Analgesics: Neurochemical, Biobehavioral and Clinical Perspectives,* Kuhar, M. and Pasternak, G., eds., Raven Press, New York, 1984, 175.

11. Kokka, N. and Fairhurst, A. S. Naloxone enhancement of acetic acid-induced writhing in rats, Life Sci., 21, 975, 1977.

12. Ramabadran, K. and Jacob, J., Stereospecific effects of opiate antagonists on superficial and deep nociception and on motor activity suggest involvement of endorphins on different opioid receptors, Life Sci., 24, 1959, 1979.

13. Chernov, H. I., Wilson, D. E., Fowler, F. and Plummer, A. J., Nonspecificity of the mouse writhing test, Arch. Int. Pharmacodyn., 167, 171, 1967.

14. Schweizer, A., Brom, R. and Scherrer, H., Combined automated writhing/motility test for testing analgesics, Agents Actions, 23, 29, 1988.

15. Taira, N., Nakayama, K. and Hashimoto, K., Vocalization of puppies to intra-arterial administration of bradykinin and other analgesic agents, Tokohu J. Exp. Med., 96, 365, 1968.

16. Guzman, F., Braun, C. and Lim, R. K. S., Visceral and pseudo affective response to intra-arterial injection of bradykinin and other agents, Arch. Int. Pharmacodyn., 136, 353, 1962.

17. Adashi, K.-I. and Ishii, Y., Vocalization to close-arterial injection of bradykinin and other analgesic agents in guinea pigs and its application to quantitative assessment of analgesic agents, J. Pharmacol. Exp. Ther., 209, 117, 1979.

18. Deffenu, G., Pergrassi, L. and Lumachi, B., J. Pharm. Pharmacol., 18, 135, 1966.

19. Martin, W. R., Pharmacology of opioids, Pharmacol., Rev., 35, 283, 1984.

20. Satoh, M., Kawagiri, S. I., Yamamoto, M., Makino, H. and Takagi, H., Reversal by naloxone of adaptation of rats to noxious stimuli, Life Sci., 24, 685, 1979.

21. Teiger, D. G., A test for the antinociceptive activity of narcotic and narcotic antagonist analgesics in the guinea pig, J. Pharmacol. Exp. Ther., 197, 311, 1976.

22. Herman, Z. S. and Felinska, W., Rapid test for screening of narcotic analgesics in mice, Pol. J. Pharmacol. Pharm., 31, 605, 1979.

23. Dubuisson, D. and Dennis, S. G., The formalin test: A quantitative study of the analgesic effects of morphine, meperidine and brain stem stimulation in rats and cats, Pain, 4, 161, 1977.

24. Hunskaar, S., Fasmer, O. B. and Hole, K., Formalin test in mice, a useful technique for evaluating mild analgesics, J. Neurosci. Meth., 14, 69, 1985.

25. Hunskaar, S. and Hole, K., The formalin test in mice: Dissociation between inflammatory pain, Pain, 30, 103, 1987.

26. Murray, C. W., Porreca, F. and Cowan, A., Methodological refinements to the mouse paw formalin test–An animal model for tonic pain, J. Pharmacol. Meth., 20, 175, 1988.

27. Alreja, M., Mutalik, P., Nayar, U. and Manchanda, S. K., The formalin test: A tonic pain model in the primate, Pain, 20, 97, 1984.

28. Wheeler-Aceto, H. and Cowan, A., Standardization of the rat paw formalin test for the evaluation of analgesics, Psychopharmacol., 104, 35, 1991.

29. Rosland, J. H., The formalin test in mice: The influence of ambient temperature, Pain, 45, 211, 1991.

30. Nielsen, P. L., Studies on analgesimetry by electrical stimulation of the mouse tail, Acta Pharmacol. (Copenh), 18, 10, 1961.

31. Romer, D., A sensitive method for measuring analgesic effects in the monkey, In *Pain,* Soulairac, A., Cahn, J. and Charpentier, J., eds., Academic Press, London, 1968, 165.

32. Huang, K. H. and Shyu, B. C., Differential stress effects on responses to noxious stimuli as measured by tail-flick latency and squeak threshold in rats, Acta Physiol. Scand., 129, 401, 1987.

33. Paalzow, G. and Paalzow, L., The effects of caffeine and theophylline on nociceptive stimulation in the rat, Acta Pharmacol. Toxicol., 32, 22, 1973.

34. Carrol, M. N. and Lim, R. K. S., Observations on the neuropharmacology of morphine and morphine-like analgesics, Arch. Int. Pharmacodyn., 125, 383, 1960.

35. Vidal, C., Girault, G. and Jacob, J., The effect of pituitary removal on pain regulation in the rat, Brain Res., 233, 53, 1982.

36. Evans, W. O., A new technique for the investigation of some analgesic drugs on a reflexive behavior in the rat, Psychopharmacologia, 2, 318, 1961.

37. Bonnet, K. A. and Peterson, K. E., A modification of the jump-flinch technique for measuring pain sensitivity in rats, Pharmacol. Biochem. Behav., 3, 47, 1975.

38. Neil, A. and Terenius, L., An improved foot-shock titration procedure in rats for centrally acting analgesics, Acta Pharmacol. Toxicol., 50, 93, 1982.

39. Kramer, E. and Bodnar, R. J., Age-related decrements in morphine analgesia: A parametric analysis, Neurobiol. Aging, 7, 185, 1986.

40. Rosenfeld, J. P. and Holzman, B. S., Differential effect of morphine on stimulation of primary versus higher order trigeminal terminals, Brain Res., 124, 367, 1977.

41. Jessell, T. M. and Iversen, L. L., Opiate analgesics inhibit substance P release from rat trigeminal nucleus, Nature, 268, 549, 1977.

42. Andersen, R. K., Lund, J. P. and Puil, E., Enkephalin and substance P effects related to trigeminal pain, Can. J. Physiol. Pharmacol., 56, 216, 1978.

43. Vierck, Jr, C. J. and Cooper, B. Y., Guidelines for assessing pain reactions and pain modulation in laboratory animal subjects, In Advances in Pain Res. Ther., Kruger, L. and Liebeskind, J. C., eds., Raven Press, New York, 1984, p. 305.

44. Cooper, B. Y. and Vierck, Jr., C. J., Vocalization as measures of pain in monkeys, Pain, 26, 309, 1986.

45. Weiss, B. and Laties, V. G., Characteristics of aversive thresholds measured by a titration schedule. J. Exp. Analysis Behav., 6, 563, 1963.

46. Dykstra, L. A. and McMillan, D. E., Electric shock titration: Effects of morphine, methadone, pentazocine, nalorphine, naloxone, diazepam and amphetamine, J. Pharmacol. Exp. Ther., 202, 660, 1977.

47. Yaksh, T. L. and Rudy, T. A., A dose ratio comparison of the interaction between morphine and cyclazocine with naloxone in rhesus monkeys on the shock titration task, Eur. J. Pharmacol., 46, 83, 1977.

48. Dykstra, L. A., Effects of morphine, pentazocine and cyclazocine alone and in combination with naloxone on electric shock titration in the squirrel monkey, J. Pharmacol. Exp. Ther., 211, 722, 1979.

49. Bloss, J. L. and Hammond, D. L., Shock titration in the rhesus monkey: Effects of opiate and nonopiate analgesics, J. Pharmacol. Exp. Ther., 235, 423, 1985.

50. Lineberry, C. G., Laboratory animals in pain research, In Methods of Animal Experimentation, Vol. VI, Gay, W. I., ed., Academic Press, New York, 1981, 237.

51. Weiss, B. and Laties, V. C., Analgesic effects in monkeys of morphine, nalorphine and a benzomorphan narcotic antagonist, J. Pharmacol. Exp. Ther., 143, 169, 1964.

52. Yim, G. K. W., Keasling, H. H., Gross, E. G. and Mitchell, C. W., Simultaneous respiratory minute volume and tooth pulp threshold changes following levorphanol, morphine and levorphan-levallorphan mixture in rabbits, J. Pharmacol. Exp. Ther., 115, 96, 1955.

53. Chin, J. H. and Domino, E. F., Effect of morphine on brain potentials evoked by stimulation of the tooth pulp of the dog, J. Pharmacol. Exp. Ther., 132, 74, 1961.

54. Mitchell, C. L., A comparison of drug effects upon the jaw jerk response to electrical stimulation of the tooth pulp in dogs and cats, J. Pharmacol. Exp. Ther., 146, 1, 1964.

55. Shigenaga, Y., Matano, S., Okada, K. and Sakai, A., The effect of tooth pulp stimulation in the thalamus and hypothalamus of the rat, Brain Res., 63, 402, 1973.

56. Chan, S. H. H. and Fung, S. J., The effect of morphine on the jaw opening reflex in rabbits, Exp. Neurol., 53, 363, 1976.

57. Ha, H., Wu, R. S., Contreras, R. A. and Tan, E. C., Measurement of pain threshold by electrical stimulation of tooth pulp afferents in the monkey, Exp. Neurol., 61, 260, 1978.

58. Skingle, M. and Tyers, M. B., Evaluation of antinociceptive activity using electrical stimulation of the tooth pulp in the conscious dog, J. Pharmacol. Method., 2, 71, 1979.

59. Skingle, M. and Tyers, M. B., Further studies on opiate receptors that mediate antinociception: Tooth pulp stimulation in dog, Brit. J. Pharmacol., 70, 323, 1980.

60. Piercey, M. F. and Schroeder, L. A., A quantitative analgesic assay in the rabbit based on the response to tooth pulp stimulation, Arch. Int. Pharmacodyn. Ther., 248, 294, 1980.

61. Toda, K. and Iriki, A., Quantitative relation between the intensity of tooth pulp stimulation and the magnitude of jaw opening reflex in the rat, Physiol. Behav., 24, 1173, 1980.

62. Wynn, R. L., El'Baghdady, Y. M., Ford, R. D., Thut, P. D. and Rudo, F. G., A rabbit tooth pulp assay to determine ED_{50} values and duration of action of analgesics, J. Pharmacol. Meth., 11, 109, 1984.

63. Steinfels, G. F. and Cook, L., Antinociceptive profiles of mu and kappa opioid agonists in rat tooth pulp stimulation procedure, J. Pharmacol. Exp. Ther., 236, 111, 1986.

64. Rajaona, J., Dallel, R. and Woda, A., Is electrical stimulation of the rat incisor an appropriate experimental nociceptive stimulus?, Exp. Neurol., 93, 291, 1986.

65. Wynn, R. L. and Bergman, S. A., Quantification of the morphine reversal activity of opioid agonist/antagonists and naloxone using a rabbit tooth pulp procedure, Drug. Dev. Res., 11, 19, 1987.

66. Matthews, B. and Searle, B. N., Electrical stimulation of teeth, Pain, 2, 245, 1976.

67. Greenwood, F., Horiuchi, H. and Matthews, B., Electrophysiological evidence on the types of nerve fibers excited by electrical stimulation of the teeth with a pulp tester, Arch. Oral Biol., 17, 701, 1972.

68. Oliveras, J. L., Woda, A., Guilbaud, G. and Besson, J. M., Inhibition of the jaw opening reflex by electrical stimulation of the periaqueductal gray matter in the awake unrestrained cat, Brain Res., 72, 328, 1974.

69. Mason, P., Strassman, A. and Maciewicz, R., Is the jaw opening reflex a valid model of pain, Brain Res. Rev., 10, 137, 1985.

70. Haffner, F., Experimentelle prufung schemerzstillender mittel, Dt. Med. Wschr., 55, 731, 1929.

71. Bianchi, C. and Franceschini, I., Experimental observations on Haffner's method for testing analgesic drugs, Brit. J. Pharmacol., 9, 280, 1954.

72. Green, A. F., Young, P. A. and Godfrey, E. I., A comparison of heat and pressure analgesiometric methods in rats, Brit. J. Pharmacol., 6, 572, 1951.

73. Sakurada, S., Sakurada, T., Jin, H., Kisara, K., Sasaki, Y. and Suzuki, K., Antinociceptive activities of synthetic dipeptides in mice, J. Pharm. Pharmacol., 34, 750, 1982.

74. Furuta, S., Kisara, K., Sakurada, S., Sakurada, T., Sasaki, Y. and Suzuki, K., Structure-antinociceptive activity studies with neurotensin, Brit. J. Pharmacol., 83, 43, 1984.

75. Vane, J. R., Inhibition of prostaglandin synthesis as a mechanism of action for aspirin-like drugs, Nature, 231, 232, 1971.

76. Ferreira, S. H., Nakamura, M. and DeAbreu Castro, M.S., The hyperalgesic effects of prostacyclin and prostaglandin E_2, Prostaglandins, 16, 31, 1978.

77. Randall, L. O. and Selitto, J. J., A method for measurement of analgesic activity on inflamed tissue, Arch Int. Pharmacodyn., 111, 409, 1957.

78. Gilfoil, T. M., Klavins, I. and Grumbach, L., Effects of acetylsalicylic acid on the edema and hyperaesthesia of the experimentally inflamed rat's paw, J. Pharmacol. Exp. Ther., 142, 1, 1963.

79. Ferreira, S. H., Lorenzetti, B. B. and Correa, F. M. A., Central and peripheral antialgesic action of aspirin-like drugs, Eur. J. Pharmacol., 53, 39, 1978.

80. Romer, D., Pharmacological evaluation of mild analgesics, Brit, J. Clin. Pharmacol., 10, 247S, 1980.

81. Winter, C. A., Kling, P. J., Tocco, D. J. and Tanabe, K., Analgesic activity of diflunisal [MK-647;5-(2,4-diflurophenyl)salicylic acid] in rats with hyperalgesia induced by Freund's adjuvant, J. Pharmacol. Exp. Ther., 211, 678, 1979.

82. Hayes, A. G., Sheehan, M. J. and Tyers, M. B., Differential sensitivity of models of antinociception in the rat, mouse and guinea-pig to μ- and K-opioid receptor agonists, Brit. J. Pharmacol., 91, 823, 1987.

83. Ramabadran, K. and Bansinath, M., The role of endogenous opioid peptides in the regulation of pain, Crit. Rev. Neurobiol., 6, 13, 1990.

84. Rios, L. and Jacob, J. J. C., Inhibition of inflammatory pain by naloxone and its N-methyl quaternary analogue, Life Sci., 31, 1209, 1982.

85. Rios, L. and Jacob, J. J. C., Local inhibition of inflammatory pain by naloxone and its N-methylquaternary analogue, Eur. J. Pharmacol., 96, 277, 1983.

86. Berry, J. J., Montgomery, L. D. and Williams, B. A., Thermoregulatory responses of rats to varying environmental temperatures, Aviat. Space Environ. Med., 55, 546, 1984.

87. Rand, R. P., Burton, A. C. and Ing, T., The tail of the rat in temperature regulation and acclimatization, Can. J. Physiol. Pharmacol., 43, 257, 1965.

88. Ness, T. J. and Gebhart, G. F., Centrifugal modulation of the rat tail flick reflex evoked by graded noxious heating of the tail, Brain Res., 386, 41, 1986.

89. Eide, P. K., Berge, O.-G., Tjolsen, A. and Hole, K., Apparent hyperalgesia in the mouse tail flick test due to increased tail skin temperature after lesioning of sertonergic pathways, Acta Physiol. Scand., 134, 413, 1988.

90. Vidal, C., Suaudeau, C. and Jacob, J., Regulation of body temperature and nociception induced by non-noxious stress in rats, Brain Res., 297, 1, 1984.

91. D'Amour, F. E. and Smith, D. L., Method for determining loss of pain sensation, J. Pharmacol. Exp. Ther., 72, 74, 1941.

92. Ramabadran, K. and Bansinath, M., Bioscreening technique for analgesic activity, In Drug Bioscreening Drug evaluation techniques in Pharmacology, E. B. Thompson, ed., VCH, New York, 1990, 53.

93. Gray, W. D., Osterberg, A. C. and Scuto, T. J., Measurement of analgesic efficacy and potency of pentazocine by D'Amour and Smith Method, J. Pharmacol. Exp. Ther., 172, 154, 1970.

94. Appelbaum, B. D. and Holtzman, S. G., Restraint stress enhances morphine induced analgesia in the rat without changing apparent affinity of receptor, Life Sci., 36, 1069, 1985.

95. Kelly, S. J. and Franklin, K. B. J., Role of peripheral and central opioid activity in analgesia induced by restraint stress, Life Sci., 41, 789, 1987.

96. Ramabadran, K. and Jen, M. F., Opioid agonist and antagonist properties of diasteroisomeric N-Tetrahydronooxymophones in mice, Arch. Int. Pharmacodyn. Ther., 274, 180, 1985.

97. Fennessy, M. R. and Lee, J. R., The assessment of and the problems involved in the experimental evaluation of narcotic analgesics, In Methods in Narcotic Research Ehrenpreis, S. and Neidle, A. eds., Marcel Dekker, Inc., New York, 1975, 73.

97a. Wright, D. M., Diurnal rhythm in sensitivity of a nociceptive spinal reflex, Eur. J. Pharmacol., 69, 385, 1981.

98. Yoburn, B. C., Morales, R., Kelly, D. D. and Inturrisi, C. E., Constraints on the tail flick assay: Morphine analgesia and tolerance are dependent upon locus of tail stimulation, Life Sci., 34, 1755, 1984.

99. Duggan, A. W., Griersmith, B. T., Headley, P. M. and Maher, J. B., The need to control skin temperature when using radiant heat in tests of analgesia, Exp. Neurol., 61, 471, 1978.

100. Jen, M. F. and Han, J., Rat tail flick acupuncture analgesia model, Chin. Med. J., 92, 576, 1979.

101. Tricklebank, M. D. and Curzon, G., In Stress induced analgesia, Tricklebank, M. D. and Curzon, G., eds, John Wiley and Sons, New York, 1984, 185.

102. Porro, C. A. and Carli, G., Immobilization and restraint effects on pain reactions in animals, Pain, 32, 289, 1988.

103. Rodgers, R. J. and Randall, J. I., Environmentally induced analgesia: Situational factors, mechanisms and significance, In Endorphins, opiates and behavioral processes, Rodgers, R. J. and Cooper, S. J., eds., John Wiley and Sons, New York, 1988, 109.

104. Dewey, W. L. and Harris, L. S., The tail flick test, In Methods in narcotic research, Ehrenpreis, S. and Neidle, A., eds., Marcel Dekker Inc., New York, 1975, 101.

105. Mayer, D. J., Wolfle, T., Akil, H., Carder, B. and Liebeskind, J. C., Analgesia from electrical stimulation in the brain stem of the rat, Science, 1351, 1971.

106. Janssen, P. A. J., Niemegeers, C. J. E. and Dony, J. G. H., The inhibitory effect of fentanyl and other morphine-like analgesics on the warm water induced tail withdrawal reflex in rats, Arznm. Forsch, 13, 502, 1963.

107. Ramabadran, K., Bansinath, M., Turndorf, H. and Puig, M. M., Tail immersion test for the evaluation of a nociceptive reaction in mice, J. Pharmacol. Meth., 21, 21, 1989.

108. Sewell, R. D. E. and Spencer, P. S. J., Antinociceptive activity of narcotic agonist and partial agonist analgesics and other agents in the tail immersion test in mice and rats, Neuropharmacology, 15, 683, 1976.

109. Luttinger, D., Determination of antinociceptive efficacy of drugs in mice using different water temperatures in a tail immersion test, J. Pharmacol. Method., 13, 351, 1985.

110. Jacob, J. J. C. and Ramabadran, K., Enhancement of a nociceptive reaction by opioid antagonists in mice, Brit. J. Pharmacol., 64, 91, 1978.

111. Ramabadran, K. and Bansinath, M., Morphine-naloxone synergism, Anesth. Analg., 65, 1252, 1986.

112. Pizziketti, R. J., Pressman, N. S., Geller, E. B., Cowan, A. and Adler, M. W., Rat cold water tail flick: A novel analgesic test that distinguishes opioid agonists from mixed agonist-antagonists, Eur. J. Pharmacol., 119, 23, 1985.

113. Clark, S. J., Follenfant, R. L. and Smith, T. W., Evaluation of opioid-induced antinociceptive effects in anesthetized and conscious animals, Brit. J. Pharmacol., 95, 275, 1988.

114. Tiseo, P. J., Geller, E. B. and Adler, M. W., Antinociceptive action of intracerebroventricularly administered dynorphin and other opioid peptides in rat, J. Pharmacol. Exp. Ther., 246, 449, 1988.

115. Woolfe, G. and MacDonald, A. D., The evaluation of the analgesic action of pethidine hydrochloride (Demerol). J. Pharmacol. Exp. Ther., 80, 300, 1944.

116. Eddy, N. B. and Leimbach, D., Synthetic analgesics. II. Diethienylbutenyl and dithienylbutylamines, J. Pharmacol. Exp. Ther., 107, 385, 1953.

117. Jacob, J. and Blozovski, M., Action de divers analgéiques sur le comportémènt de souris exposées a un stimulus thérmoanalgésiques (II): Apprentissage nocicéptif d'urgènce. Actions differentielles de substances analgésiques et psyhoreac- tives sur lés réactions de léchmènt et le bond, Arch. Int. Pharmacodyn., 133, 296, 1961.

118. Jacob, J., Lafille, C., Loiseau, G., Echinard-Garin, P. and Barthélemy, C., Récherches concernant la caractérisation et al différentiation pharmacologiques des drogues hallucinogénès (dérives indoliques et mescaline, nalorphine, anticholinérgiques centraux, phéncyclidine), L'encephale, 4, 1, 1964.

119. Jacob, J., Loiseau, G., Echinard-Garin, P., Barthelemy, C. and Lagille, C., Caracterisation et detection pharmacologiques des substances hallucinogines (II): Antagonism vis-á-vis de la morphine chez la souris, Arch. Int. Pharmacodyn., 148, 14, 1964.

120. Jacob, J., Tremblay, E. C. and Colombel, M. C., Facilitation de reactions nociceptives par la naloxone chez la souris et le rat, Psychopharmacologia (Berlin), 37, 217, 1974.

121. Jacob, J. J. C. and Ramabadran, K. Opioid antagonists, endogenous ligands and nociception, Eur. J. Pharmacol., 46, 393, 1977.

122. Frederickson, R. C. A., Burgis, V. and Edwards, J. D., Hyperalgesia induced by naloxone follows diurnal rhythm in responsivity to painful stimuli, Science, 198, 756, 1977.

123. Grevert, P. and Goldstein, A., Some effects of naloxone on behavior in the mouse, Psychopharmacologia, 53, 111, 1977.

124. Ramabadran, K., Bansinath, M., Turndorf, H. and Puig, M. M., The hyperalgesic effect of naloxone is attenuated in streptozotocin-diabetic mice, Psychopharmacology, 97, 169, 1989.

125. Jacob, J., Ramabadran, K., Girault, J. M., Suaudeau, C. and Michaud, G., Endorphins, training and behavioral thermoregulation, In Characteristics and functions of opioids, Dev. Neurosci, Vol. 4., Van Ree, J. M., Terenius, L., eds., North Holland Biomedical Press, Amsterdam, 1978, 171.

126. Ramabadran, K., Guillon, J. C. and Jacob, J., Action de la naloxone sur l'apprentissagenociceptif (Théophyllineetméthoxy-5, dimèthyl-N,N-tryptamine), J. Pharmacol. (Paris), 10, 264, 1979.

127. Ramabadran, K., Guillon, J. C. and Jacob, J., Action of hyperalgesic substances [(-)naloxone, theophylline and 5-methoxy-N,N-dimethyl-tryptamine] on nociceptive learning, In Exogenous and endogenous opiate agonists and antagonists, Way, E. L., ed. Pergamon Press, New York, 1980, 471.

128. Lai, Y.-Y. and Chan, S. H. H., Shortened pain response time following repeated algesiometric tests in rats, Physiol. Behav., 28, 1111, 1982.

129. Messing, R. B., Rijk, H. and Rigter, H., Facilitation of hot plate response learning by pre-and posttraining naltrexone administration, Psychopharmacology, 81, 33, 1983.

130. Rodgers, R. J., Randall, J. and Pittock, F., Hot plate learning in mice is unaltered by immediate post-training administration of naloxone, naltrexone or morphine, Neuropharmacology, 24, 333, 1985.

131. Gamble, G. D. and Milne, R. J., Repeated exposure to sham testing procedures reduces reflex withdrawal and hot plate latencies: Attenuation of tonic descending inhibition?, Neurosci. Lett., 96, 312, 1989.

132. O'Callaghan, J. P. and Holtzman, S. G., Quantification of the analgesic activity of narcotic antagonists by a modified hot plate procedure, J. Pharmacol. Exp. Ther., 192, 497, 1975.

133. Ramabadran, K., Michaud, G. and Jacob, J. J. C., Genetic influences on the control of nociceptive responses and precipitated abstinence, Ind. J. Exp. Biol., 20, 74, 1982.

134. Ramabadran, K., Strain differences in the hyperalgesic effect of a novel k receptor agonist, U-50488H in mice, Eur. J. Pharmacol., 98, 425, 1984.

135. Ramabadran, K., Stereospecific enhancement of nociception by opioids in different strains of mice, Jpn. J. Pharmacol., 37, 296, 1985.

136. Bansinath, M., Ramabadran, K., Turndorf, H. and Puig, M. M., Effects of the benzomorphan *k* opiate, MR 2266 and its (+) enantiomer MR 2267, on thermonociceptive reactions in different strains of mice, Neurosci. Lett., 117, 212, 1990.

137. Ramabadran, K., Bansinath, M., Turndorf, H. and Puig, M. M., Effect of naloxone and Mr 2266 on thermonociceptive reactions in diabetic mice, NIDA Res. Monograph, 95, 528, 1990.

138. Bansinath, M., Ramabadran, K., Turndorf, H. and Puig, M. M., *k*-opioid receptor-mediated thermonociceptive mechanisms in streptozotocin diabetes, Physiol. Behav., 49, 729, 1991.

139. Pick, C. G., Cheng, J., Paul, D. and Pasternak, G., Genetic influences in opioid analgesic sensitivity in mice, Brain Res., 566, 295, 1991.

140. Yanaura, S., Yamatake, Y. and Ouchi, T., A new analgesic testing method using ultrasonic stimulation, Jpn. J. Pharmacol., 26, 301, 1976.

141. Mohrland, J. S., Johnson, E. E. and VonVoigtlander, P. F., An ultrasound-induced tail flick procedure: Evaluation of nonsteroidal antiinflammatory analgesics, J. Pharmacol. Meth., 9, 279, 1983.

142. Carmon, A. and Frosig, R., Noxious stimulation of animals by brief intense laser induced heat: Advantages to pharmacological testing of analgesics, Life Sci., 29, 11, 1981.

143. Basbaum, A. I., Effects of central lesions on disorders produced by multiple dorsal rhizotomy in rats, Exp. Neurol., 42, 490, 1974.

144. Duckrow, R. B. and Taub, A., Effect of diphenylhydantoin on self-mutilation in rats produced by unilateral multiple dorsal rhizotomy, Exp. Neurol., 54, 33, 1977.

145. Lombard, M. C., Nashold, B. S., Albe-Fessard, D., Salman, N. and Sakr, C., Deafferentation hypersensitivity in the rat after dorsal rhizotomy: A possible animal model of chronic pain, Pain, 6, 163, 1979.

146. Dennis, S. G. and Melzack, R., Self mutilation after dorsal rhizotomy in rats: Effects of prior pain and pattern of root lesions, Exp. Neurol., 65, 412, 1979.

147. Wiesenfeld, Z. and Lindblom, U., Behavioral and electrophysiological effects of various types of peripheral nerve lesions in the rat: A comparison of possible models for chronic pain, Pain, 8, 283, 1980.

148. Levitt, M. and Heybach, J. P., The deafferentation syndrome in genetically blind rats: A model of the painful phantom limb, Pain, 10, 67, 1981.

149. Coderre, T. J., Grimes, R. W. and Melzack, R., Deafferentation and chronic pain in animals: An evaluation of evidence suggesting autotomy is related to pain, Pain, 26, 61, 1986.

150. Albe-Fessard, D. and Lombard, M. C., Use of animal model to evaluate the origin of and protection against deafferentation pain, In Adv. Pain Res. Ther., 5, Bonica, J. J., Lindblom, U. and Iggo, A., eds., Raven Press, New York, 1983, 691.

151. Govrin-Lippmann, R. and Devor, M., Ongoing activity in severed nerves: Source and variation with time, Brain Res., 159, 406, 1978.

152. Wall, P. D., Devor, M., Inbal, R., Scadding, J. W., Schonfield, D., Seltzer, Z. and Tomkiewicz, M. M., Autotomy following peripheral nerve lesions: Experimental anesthesia dolorosa, Pain, 7, 109, 1979.

153. Kingery, W. S. and Vallin, J. A., The development of chronic mechanical hyperalgesia, autotomy and collateral sprouting following sciatic nerve section in rat, Pain, 38, 321, 1989.

154. Xu, X.-J. and Wiesenfeld-Hallin, Z., The threshold for the depressive effect of intrathecal morphine on the spinal nociceptive flexor reflex is increased during autotomy after sciatic nerve section in rats, Pain, 46, 223, 1991.

155. Rodin, B. E. and Kruger, L., Deafferentation in animals as a model for the study of pain: An alternative hypothesis, Brain Res. Rev., 7, 213, 1984.

156. Wiesenfeld-Hallin, Z., The effects of intrathecal morphine and naltrexone on autotomy in sciatic nerve sectioned rats, Pain, 18, 267, 1984.

157. Wiesenfeld, Z. and Hallin, R. G., Continuous naloxone administration via osmotic minipump decreases autotomy but has no effect on nociceptive threshold in the rat, Pain, 16, 145, 1983.

158. Suaudeau, C., DeBeayrepaire, R., Rampin, O. and Albe-Fessard, D., Antibiotics and morphinomimetic injections prevent automutilation behavior in rats after dorsal rhizotomy, Clin. J. Pain, 5, 177, 1989.

159. Bennett, G. J. and Xie, Y., A peripheral mononeuropathy in rat that produces disorders of pain sensation like those seen in man, Pain, 33, 87, 1988.

160. Hirose, K. and Jyoyama, H., Measurement of arthritic pain and effects of analgesics in the adjuvant treated rat, Jpn. J. Pharmacol., 21, 717, 1971.

161. Colpaert, F. C., Evidence that adjuvant arthritis in the rat is associated with chronic pain, Pain, 28, 201, 1987.

162. Kuzuna, S. and Kawai, K., Evaluation of analgesic agents in rats with adjuvant arthritis, Chem. Pharm. Bull., 23, 1184, 1975.

163. Pircio, A. W., Fedele, C. T. and Bierwagen, M. E., A new method for the evaluation of analgesic activity using adjuvant induced arthritis in the rat, Eur. J. Pharmacol., 31, 207, 1975.

164. DeCastro Casta, M., DeSutter, P., Gybels, J. and Van Hees, J., Adjuvant induced arthritis in rats: A possible animal model for chronic pain, Pain, 10, 173, 1981.

165. Coderre, T. J. and Wall, P. D., Ankle joint urate arthritis (AJUA) in rats: An alternative model of arthritis to that produced by Freund's adjuvant, Pain, 28, 379, 1987.

166. Okuda, K., Nakahama, H., Miyakawa, H. and Shima, K., Arthritis induced in cats by sodium urate: A possible animal model for chronic pain, Pain, 18, 287, 1984.

167. Dubner, R., Methods of assessing pain in animals, In Textbook of pain, Wall, P. D. and Melzack, R., eds., Churchill Livingstone, London, 1989, 247.

168. Ramabadran, K. and Jacob, J. J. C., Effect of various sertonergic agonists and an antagonist on a nociceptive reaction in mice, Jpn. J. Pharmacol., 32, 1059, 1982.

169. Jacob, J. J, Barthelemy, C. D., Tremblay, E. C. and Colombel, M.C., Potential usefulness of a single dose acute physical dependence on and tolerance to morphine for the evaluation of narcotic antagonists, In Narcotic antagonists, Adv. Biochem. Psychopharmacol., Vol. 8, Braude, M. C., Harris, L. S., May, E. L., Smith, J. P. and Villarreal, J. E., eds., Raven Press, New York, 1974, 299.

170. Ramabadran, K., Naloxone-precipitated abstinence in mice, rats and gerbil acutely dependent on morphine, Life Sci., 33, Suppl. I., 385, 1983.

171. Ramabadran, K., An analysis of precipitated withdrawal in rats acutely dependent on morphine, Jpn. J. Pharmacol., 37, 307, 1985.

Index

Abu Abdullah Mohammed, 4
Abu-Ali-Ibn-Sina (Avicenna), 3
Acetate metabolism, relationship
    to alkaloid biosynthesis, 251
Acetic acid, as chemical writhing
    agent, 261
14B-Acetoxydihydrocodeinone,
    176,178
Acid, organic, laticifer content, 113
Acyrthosiphon, 86
Adam (biblical figure), 7
Aeneid (Virgil), 7
Afghanistan, *Papaver somniferum*
    cultivation in, 24,41
Africa, *Papaver somniferum*
    cultivation in, 24
Agaric, bulbous, 246
Agaricaceae, 245
Agricultural studies, of *Papaver*
    somniferum, 65-94
    diseases and control measures,
      80-85
      capsule infection, 83-84
      downy mildew, 81
      leaf blight, 82-83
      powdery mildew, 81-82
      root rot, 84
      seed-borne diseases, 82
      seedling blight, 82
      viral diseases, 84-85
      wilt, 84
    fertilizers and manures, 66-69
    harvesting, 71-75
    irrigation, 69-70
    morphine concentration changes,
      80
    pests, 85-88
    seedling thinning, 66

Agricultural studies, of *Papaver*
    somniferum (continued)
    soil requirements, 65
    sowing, 65-66
    weather hazards, 85
    weeding and hoeing, 70-71
    yield of opium/morphine, 75-80
Agrobacterium rhizogenes, 59
Agrotis suffusa, 85-86
Akbar, 11
Alachlor, 71
Alborine, 38
Alcohol, interaction with morphine,
    174
Aldrin, 69
Alexander the Great, 7,10
Algae, alkaloid occurrence in, 245
Alkaloidal ring systems, phenol
    radicals in, 218
Alkaloid-producing plants
    animals' avoidance of, 245-246
    animals' consumption of, 246
    definition, 244
    geographic distribution, 243
Alkaloids. *See also* Opium alkaloids;
    names of specific alkaloids
    biosynthesis, xiv, 250-251
    biosynthesis, site of, 192-195
    bitterness, 246
    early research in, 241-242
    functions, 245-251
    isolation, 241-242
    laticifer content, 113
    metabolic status, 247-248
    number of, 242
    occurrence, 242-245
    during plants' developmental
      stages, 249

299